MW01288846

Earth Is A Nice Little Planet:

Who Will Save It?

Christina Park
&
Orville Powell

AuthorHouse™
1663 Liberty Drive, Suite 200
Bloomington, IN 47403
www.authorhouse.com
Phone: 1-800-839-8640

First published by AuthorHouse 8/28/2007

ISBN: 978-1-4343-2349-1 (sc)

Printed in the United States of America
Bloomington, Indiana

This book is printed on acid-free paper.

Dedication

This book is dedicated to Mom and Dad -
for strong roots and far-reaching wings.
And to Orv - for a real challenge. Thank you!

~ Christina ~

This book is dedicated to my grandchildren,
Christine Bertolette, Mason Powell,
Campbell Powell, Marek Emilio Powell
and to my great granddaughter, Caylee Bertolette.
It is their future that is at stake.

~ Orville ~

Contents

Dedication v

Acknowledgments ix

Introduction:
A Call to Leadership xiii

Chapter One:
World Population Growth 1

Chapter Two:
United States Population Growth 19

Chapter Three:
Global Warming 35

Chapter Four:
World Food Supply 55

Chapter Five:
United States Food Supply 67

Chapter Six:
The World's Fresh Water 77

Chapter Seven:
United States Fresh Water Supply 89

Chapter Eight:
Alarm bells are ringing! 99

Chapter Nine:
A Summation 109

Chapter Ten:
The Final Word 113

References 117

Acknowledgments

The authors owe five very special people a deep sense of gratitude for making this book possible: Susan Ballard, Penny Hudoff, Steve and Ingrid Park and Dianne Powell. Their encouragement, wisdom and advice coupled with the massive amount of time they gave to editing and proof-reading drafts made this a better book. Any flaws that remain are solely the authors.

THIS BOOK WAS WRITTEN NOT TO SCARE, BUT TO INFORM.

INTRODUCTION:
A Call to Leadership

This book came about purely by coincidence. I began researching the topic of population growth when my co-author needed a new challenge during her graduate studies. I had been thinking for some time about the issue and what kinds of new challenges it will bring to local governments in the coming years. The result was our decision to jointly write this little book, in the hopes of "connecting the dots" as we like to call it. We wanted to try and find the links between such problems as increasing population growth, urban sprawl, and global warming, and then put them into a format useable for communities across the nation. We agreed that this should be a large enough challenge for my co-author and hopefully address my concerns about the impacts on local government.

What began as a small research project soon grew into something greater than we conceived. We soon realized that a lack of information on the subject was not our problem. Our problem was how to make sense of it all in a way that would be meaningful for local leaders. What we found in our research was a little daunting. We were amazed by the population growth projections and their far reaching impacts, which have the potential to cause serious problems for local communities if not properly planned for. Just as scary, is that despite the wide range of literature on the subjects addressed here, the public seems to be only marginally aware and interested in these issues. Our nation's leaders also seem unwilling to discuss these problems with us, which makes it even more difficult for the public to stay informed.

People are amazingly resilient when faced with major emergencies and natural disasters. We jump into problem solving mode with huge amounts of energy, compassion, and resources. We always expect we can "fix" the problem, and we usually do. We do not, however, give the same kind of attention to future problems, even when they are predictable. And why should we worry about challenges we may face fifty years into the future? In many ways, all levels of government have willingly or have been forced to operate only in the short run. There are important reasons, however, why we need to begin thinking fifty years out. We are going to face challenges that we have yet to come across in this country. This means we have no prior experience to fall back on and we need to begin planning today for tomorrow.

The obstacles we are talking about are not only unprecedented population growth, but global warming and the continued loss of open space, which are all irreversible. These problems can not be "fixed," rather they must be managed. We must learn and adopt effective management strategies, which will take strong leadership at all levels of government. Even more important is that if we start early enough, we can handle these challenges without resorting to limiting personal freedoms or delegating all of the responsibility to future generations.

So why are these problems going to be so difficult to manage? We have the ability to collect and analyze information and use it to make predictions about the future. We have the intellect to make appropriate management choices to shape the environment we live in and to avoid future disasters. Despite our capabilities, managing these challenges will not be an easy task for the following reasons:

- The first step in correcting any problem is to accept that there is a problem. Individuals often only see the world from a narrow perspective within their own community.

They may notice their city growing, more cars on the road, and higher temperatures, but they do not necessarily connect what is happening at home to what is happening around the world. Until we start making that connection we won't be able to understand the magnitude of these issues.

- The approaching challenges do not meet what we call the "CNN© test." This means that people cannot see there really is a problem. Scientists tell us that these problems are real, but until we find a way to "show" it to them, people won't believe that they are. The changes we are experiencing are slow and constant; the kind of changes that can sneak up on us and go unnoticed for years. As we discussed above, people can see their communities growing, but that doesn't mean they connect that to national and even worldwide growth.

- The political system in the U.S. is flawed and under the influence of powerful special interest organizations. These groups have self-serving political, religious, and financial agendas that are often threatened by policy changes. They continue to wield their influence regardless of the benefits that changes may have on society as a whole.

- Our absolute faith in technology to solve these problems is an issue. Given the magnitude of these future threats and the high stakes involved, it is risky to rely solely on technology. We have no idea what kinds of technology will be available to communities in fifty years, so we need to start coming up with alternative strategies.

- Many individuals have a hard time making sacrifices for the "good of society," especially when the benefits won't be seen for years to come. Americans have become accustomed to a certain standard of living, made possible in many ways by being a wasteful culture. We have huge homes, drive gas-guzzling SUVs, and continually buy new things rather than fix what we have. We need to change our concept of what the "American Dream" really means in order to maintain our standard of living. The sacrifices we need to make are preventative in nature and it will be difficult in the future to communicate to the public what their collective efforts have prevented.

- While scientists and demographers have been trying to alert us to these challenges, many of the world's leaders, as well as the media, have given little attention to these issues. The attention they have given has often been to question scientific findings and cast doubt in everyone's minds, making it difficult for decision-makers to agree on a course of action. There have been some exceptions to this, such as Prime Minister Tony Blair and President Vladimir Putin, who have been speaking for several years about global warming and population growth.

- Even as the news media has become more active in reporting on global warming, they never make the connection to population growth as a driving factor.

Despite the number and enormity of problems facing cities today, we the authors are still eternal optimists. We believe that while we humans are the cause of many of these problems, we are also problem solvers at heart when given the right information. The key, however, is that if we wait too long to identify and recognize the threats we face, will end up reacting too slowly to make a difference.

This book is a call to leadership. It was written for the local leaders of communities across the nation. For elected officials, local business owners, community leaders, government employees, and everyone else who has a role in shaping our cities. One longtime Speaker of the House, Thomas O'Neill, put it best when he said "all politics is local." In the end the problems of the world are really the problems of individual communities. It is time for these individuals to provide strong, courageous leadership by putting our nation on the path to solving and managing these challenges. If these leaders are to set us on the correct path, however, they must have the necessary knowledge.

This book is a compilation of a lot of that information, condensed from massive amounts of research and reports on population growth and its far reaching effects. It discusses topics such as transportation, the supply of fresh water, and energy conservation, to name a few. There are numerous books and resources on each of these topics individually and we are not breaking new ground here. We realize that local leaders have extensive demands on their time and we have therefore tried to merely provide a kind of "executive summary" of the issues. We will briefly address the topics on a global scale and then quickly relate it back to the situation in American communities. In each following chapter we will lay out the relevant issues topically and then focus on their consequences if left unchecked. We will also offer recommendations for actions that can be taken by local governments to start dealing with these issues today. At the end of the book, there is also a list of additional resources that local communities may find useful in their individual quests to tackle the problems outlined here. By acting now, generations to come will inherit a world similar to the one we enjoy today. Earth is a nice little plant. Who will save it?

CHAPTER ONE:
World Population Growth

Question:
Is the expanding world population mankind's greatest problem?

Answer:
It is certainly the most challenging for world leaders to solve. Most politicians and religious or business leaders do not understand the nature of exponential population growth. Why should they? The world has never faced this situation before and while there are tons of information available to the public, it is too numerous and too scattered for all but those working in the field to attempt to comprehend.

BACKGROUND

On February 25, 2006, the earth's population grew to an unprecedented 6.5 billion and is predicted to grow to 7 billion by October of 2012. The global population is continuing to increase at unfathomable rates. When the earth hit 6 billion in June of 1999, the U.S. Census Bureau noted that "the time required for the global population to grow from 5 billion to 6 billion – just a dozen years – was shorter than the interval between any of the previous billions."[1] By the year 2050, projections show that the world population will have grown to 9.15 billion, with only 1.23 billion people living in today's developed nations.

These projections should cause policy-makers and public administrators around the world to sit up and take notice. The sheer magnitude and speed of this population growth will present enormous challenges for mankind. Its effects will be far-reaching. Since nearly all of the services provided by governments around the world are driven by human needs, population growth will be the key factor in issues such as urban sustainability, environmental degradation, and renewable energy production, to name a few. In the end, these issues will make it increasingly difficult for governments to provide for the basic needs of their citizens.

The human race is responsible, in many ways, for all of these challenges, but we also have the ability to solve and manage them. We will not, however, be able to solve them without strong political leadership and continued education. Scientists and demographers have for decades forecasted the problems civilization would face due to population growth. In the eighteenth century, English economist and demographer Thomas Malthus predicted that the world's population would grow substantially faster than our food supply, leading to mass starvation.[2] In 1968, Paul Ehrlich wrote *The Population Bomb*, which predicted mass starvation by the end of the 1980s. Population increases occurred basically as Malthus and Ehrlich predicted, but starvation did not occur thanks to tremendous improvements in food production systems by farmers during the past forty years.[3]

So are these problems really new ones? Should we be surprised by these predictions? Should we just rely on improvements to technology to solve the problems of the future? Will technological advances provide enough food, water, health care, and energy to provide for an additional 2.5 billion people in the next 45 years? All of our research suggests that we are going to see unparalleled growth and therefore challenges on a scale we can not easily comprehend. These challenges are not new, but will become more problematic in the years to come.

It seems logical to assume that the earth has a maximum human population it can support, just as scientists discuss carrying capacities for other species. So what is that population? How close are we to reaching it? This is one point that scientists seem unable to agree upon. They have predicted populations that we have already surpassed and others we are far from reaching. Should we be worrying about this today? Are we going to have to deal with this issue within our lifetime? Edward O. Wilson, Professor Emeritus of Biology at Harvard University, said it correctly when he stated "humanity, in the desperate attempt to fit 8 billion or more people on the planet and give them a higher standard of living, is at risk of pushing the rest of life off the globe." In the end, this issue will affect all life on earth and generations to come. We are absolutely going to see the effects of population growth within our lifetime. There is no better time than the present to start planning for these issues.

Basic Human Survival

Human beings need three things to survive: air, water and food. Let's assume that the Earth has as much air now as it always had in the past thus, as long as we keep it clean enough to breathe, we will have a sufficient supply to support many billions more people than are on Earth today. So as far as amount of air available, we should not have a problem. Our task is to keep it clean and breathable.

Water continuously recycles itself so if we don't pollute it beyond human tolerances to drink, we should continue to have sufficient amounts of water on Earth to support human life years into the future. The big problem is that the fresh water is not always where we need it to be. There is a great variation as to the amount and location where rain falls on the Earth. We also dam and divert large amounts of water for irrigation, industrial use, or to supply urbanized areas. If we have the quantity of water needed, and we may not due to the melting of ice in Greenland and Antarctica, we still have huge problems in protecting its purity and its distribution. We will come back to air and water in a later chapter.

Food, the third leg of our basic survival requirement, is another manner entirely. To produce food we need suitable soils, sufficient water, and the proper climatic conditions. This, at least at first glance, appears to be the biggest limiting factor in providing for future populations. Let's look at the facts.

The Earth's surface is made up as follows:

- Total Surface Area: 197 million square miles
    - Water Surface Area: 140 million square miles
    - Land Surface Area: 57 million square miles or 36.48 billion acres.[4]

Of the total 36.48 billion acres of land surface the breakdown of this acreage for the suitability of producing food is as follows:

- 11 percent of land is suitable for agriculture (22 percent if appropriate irrigation and drainage can be provided.)
- 6 percent is permafrost
- 10 percent is too wet
- 22 percent is too shallow
- 23 percent has with chemical problems
- 28 percent is too dry [5]

The land breakdown above is from a 1992 United Nations report on population growth. The report makes two assumptions regarding the human carrying capacity of Earth. These assumptions are:

- Only 11 percent of the land surface will remain suitable for agriculture, and that
- 0.625 acres of suitable land is required to support each person

If these two assumptions are correct, then the world can support and sustain about 6.4 billion people. *This is today's population*! To support more then this number of people requires the use of renewable resources faster than they can be replaced. At the same time we speed the exhaustion of non-renewable resources.

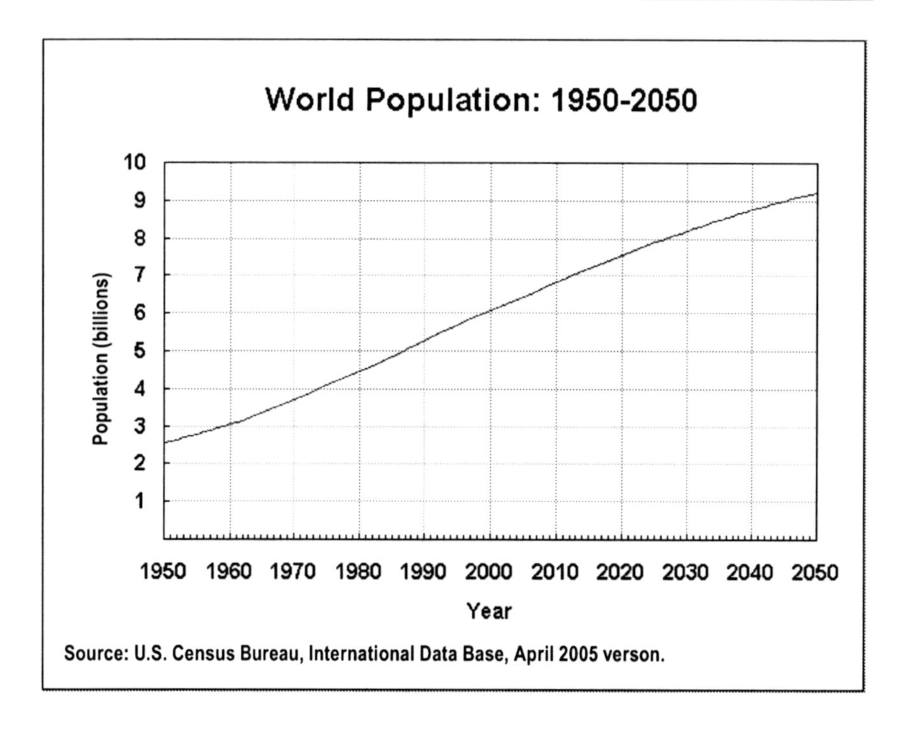

But does 6.5 billion people represent a realistic assessment of the sustainable carrying capacity of Earth? We do not know. Let us just assume that the United Nations projections are way off and that the earth can really accommodate 12 to 13 billion people. How many years will it take for the world to reach this population? At the present rate of population growth in the world, it will take approximately 40 years, or sometime around year 2046. However, as you can see by the above chart, the United States Census Bureau projects a world population of just over 9 billion people by 2050. So, if we use the lower of the two projections, sometime within the next 60 to 75 years we should reach a world population of 12 billion.

Human populations have always been restrained from their full potential to reproduce by wars, natural disasters, and epidemics. For instance, World War II claimed the lives of 59 million people, the Black Death between 1347 and 1350 resulted in the deaths of one-third of the population of Europe, and influenza from 1918 to 1919 caused the deaths of an estimated 40 to 50 million world-wide. AIDS, discovered in humans just 25 years ago, has already claimed 25 million lives. Thousands die each year from hurricanes, floods, earthquakes, and droughts. And who could forget the South Asian Tsunami of 2005 that killed 184,000 people and left 42,000 missing?[6]

There are several events that could slow today's rate of world population growth. These include:

- natural disasters such as a large meteorite striking the Earth,
- a pandemic like the H5N1 Avian Influenza (the Bird Flu) or some other unknown health threat,
- major changes in weather patterns which cause increasingly stronger storms, changes in rainfall patterns, or increased heat,
- manmade disasters such as conventional or nuclear wars,
- voluntary or mandatory birth control programs, or
- withholding medical care to certain groups.

If any of these things happened in the near future, the 2050 world population could be well below the 9 billion estimates.

However, using the 9 billion figure as being a reasonable estimate of the world's population in 2050 means that we will add 2.5 billion people to the Earth between 2006 and 2050. This is the equivalent of what the entire world's population was around 1930!

Two and a half billion people would be hard enough to plan for if all the countries were going to equally share in the growth. Of course, this is not the case. Developed nations are growing at a slower rate than the undeveloped nations and the population increases are generally predicted to take place in areas least able to handle it.

The Reuters Foundation provides us with some interesting facts on the world's population growth and where it will take place.[7]

- The population of developed countries is expected to remain approximately the same at about 1.2 billion.
- The population in the least developed countries is expected to double from 0.8 billion to 1.7 billion by 2050.

- By 2050, the population is projected to triple in Afghanistan, Burkina Faso, Burundi, Chad, Congo, Republic of Congo, East Timor, Guinea-Bissau, Liberia, Mali, Niger, and Uganda. These are not household names but they are all among the poorest countries on Earth.
- The population of 51 countries including Germany, Italy, Russia, and Japan are expected to decline through 2050.

These projections are shown in Table 1-2 below . This table compares the total population in developing countries to those in developed countries. The chart also shows how the percent of the world's population will increase in the developing countries and decrease in the developed ones.

Table 1.2: Population Growth in Developed and Developing Countries

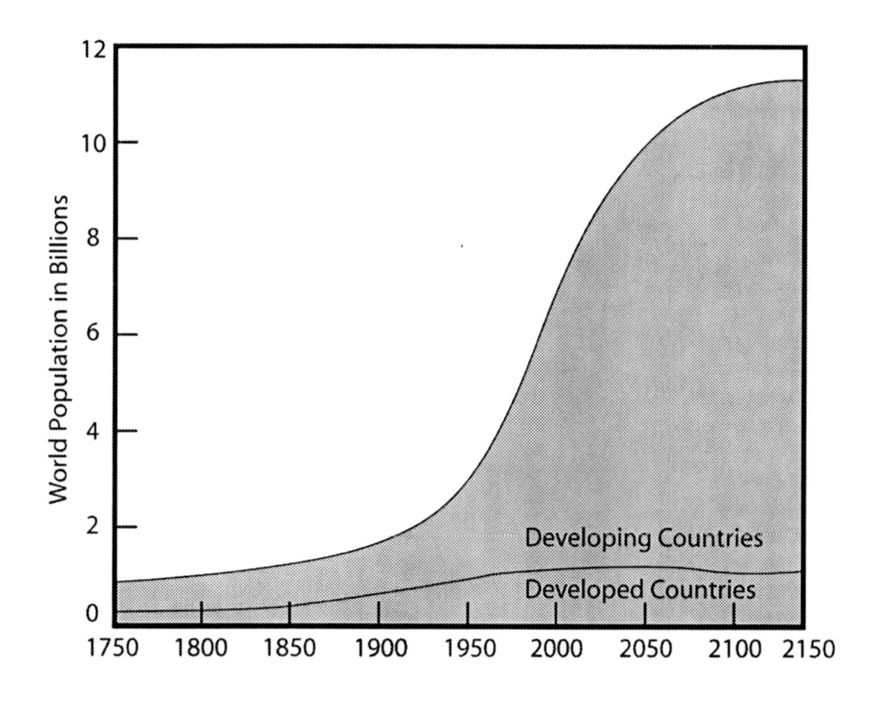

Table 1-3 breaks down the projected world population increases in billions of people and their percentage of the total for developing and developed countries.

Table 1-3: World Population Projections (2000 – 2050)[8]

		Developing Countries		Developed Countries	
Year	Total World Population (in billions)	Population (in billions)	Percent of Total Population	Population (in billions)	Percent of Total Population
2000	6.05	4.87	80.5	1.18	19.5
2010	6.86	5.62	81.9	1.24	18.1
2020	7.61	6.35	83.4	1.26	16.6
2030	8.27	7.00	84.6	1.27	15.4
2040	8.81	7.53	85.5	1.27	14.5
2050	9.16	7.92	86.5	1.23	13.5

Tables 1-4 below further depicts the world's population by comparing 2002 populations to projected 2050 statistics in the most populous nations.

Table 1-4: Most Populous Nations (2002 and 2050)[9]

2002		2050	
Country	Population (in millions)	Country	Population (in millions)
China	1,281	India	1,628
India	1,050	China	1,394
USA	287	USA	414
Indonesia	217	Pakistan	332
Brazil	174	Indonesia	316
Russia	144	Nigeria	304
Pakistan	144	Brazil	247
Bangladesh	134	Bangladesh	205
Nigeria	130	Republic of Congo	182
Japan	127	Ethiopia	173
Mexico	102	Mexico	151
Germany	82	Philippines	146
Philippines	80	Vietnam	117
Vietnam	80	Egypt	115
Egypt	71	Russia	102
Ethiopia	68	Japan	101
Turkey	67	Turkey	97
Iran	66	Iran	97
Thailand	63	Thailand	72
United Kingdom	60	Germany	68
France	60	United Kingdom	65
Italy	58	France	65
Republic of Congo	55	Italy	52

Developed nations like the United States, England, Germany, Russia, Japan, and France are experiencing declining birth rates, an aging population, increased life expectancy, and a large percentage of the population (the baby boomers) nearing retirement age. In short, these nations will start experiencing a shortage of working age adults to fill their economy's needs. These developed countries will have to rely on increasing birth rates or on migration as sources of population growth. Russia's population, for example, is declining by about 700,000 people per year and has dropped from 150 million in 1992 to 142 million today. If left unchecked, demographers estimate that Russia's population could fall to fewer than 100 million by 2050.

In response, President Vladimir Putin has asked Russian women to have more children and has suggested a "Cash-For-Babies" scheme to financially assist mothers who bear additional children. He also wants to encourage "qualified migrants" who are educated and who will "obey the law" to come and settle in Russia.[10]

Two countries with declining populations are already ahead of President Putin in encouraging women to have additional children. Germany provides a baby bonus of $2,375 a month for one year and France provides $990 a month for a year, starting with the third child.

The Elderly

To further complicate matters, the elderly population is growing at a faster rate than the younger generations. By 2050, global life expectancy is estimated to reach 75 years, up from 65 years today. Both developed and developing nations will experience increased life expectancies as follows:

- Developed nations will experience an increase from 76 to 82 years and
- Developing nations will experience an increase from 51 to 67 years.

In a publication on world population aging, the United Nations reported that "the number of older persons has tripled over the last 50 years ... [and] will more than triple again over the next 50 years."[11] The older population is also growing faster than the total population in practically all regions of the world – and the difference in growth rates is increasing. Tables 1-4 and 1-5 below illustrate the effects of aging populations on the Earth.

Table 1-5: Population Aging – The Elderly

		Number of Countries with over:		
Year	Total Population over 60 Years	10 Million Elderly	20 Million Elderly	50 Million Elderly
1950	205 million	3	3	0
2000	606 million	12	5	2
2050	Est. 2 billion	33	?	5

Table 1-6: Countries with Large Elderly Populations

1950	2000	2050
Over 10 Million Elderly	Over 20 Million Elderly	Over 50 Million Elderly
China (42 Million)	China (129 Million)	China (437 Million)
India (20 Million)	India (77 Million)	India (324 Million)
USA (20 Million)	USA (46 Million)	USA (107 Million)
	Japan (30 Million)	Indonesia (70 Million)
	Russia (27 Million)	Brazil (58 Million)

According to a recent report by MSNBC©, "the young-old balance is shifting throughout the world. In the more developed regions, the proportion of older persons already exceeds that of children, and by 2050 it is expected to be double that of children. In the less developed regions, age distribution changes have been slow but will accelerate over the next 50 years."[12] This rapidly increasing imbalance between the young and the old has one other twist. Because the mortality rates are usually higher among men than among women, females will represent a greater percentage of the elderly. This will require policy makers to address the issues of the elderly primarily as the issues of older women.

As we all know, aging brings on both physical and psychological changes. Along with psychological issues of depression, loneliness, and boredom, the elderly also face deteriorating physical conditions including poor eyesight, hearing loss, weakened muscles, brittle bones, and arthritis. The elderly also face greater risks associated with heart disorders, strokes, and pneumonia. These changes will significantly impact healthcare costs throughout the world. Societies, particularly in developed countries, will need to determine how best to meet the long-term financial needs of this increasing elderly, largely female, population. In all likelihood, there will be a much greater dependence on the children of the elderly to provide for their care.

Immigration

A decreasing and aging population can be a serious problem for countries because of the imbalance it causes in the workforce. However, countries with stable or declining populations can be a part of the solution by recruiting immigrants to supplement their populations. Nations will need to allow greater numbers of immigrants across their borders to meet workplace demands for skilled and unskilled positions.[13] This migration trend from developing to developed nations is already in full swing. According to the latest United Nations statistics, the number of international migrants reached 191 million in 2005. The United States was the preferred destination with approximately 20 percent coming to our shores; followed in popularity by Russia, Germany, Ukraine, and France.[14]

The problem in developing nations, with few exceptions, is the complete opposite. These countries are growing at alarming rates, but they do not have the resources to accommodate this growth. The countries where this growth is taking place are already struggling to feed and provide for the other basic needs of their citizens. Immigration from developing to developed nations can therefore help to solve problems for both groups. This will, however, cause serious debate as to how to allow this to occur. Who will be permitted to migrate? How will these individuals be accounted and provided for? In many developed nations the debate continues as to how to provide a certain level of service to future citizens when resources are already stretched with current populations.

Where will all these people live?

Three billion people today, nearly 50 percent of the world's population, live in cities and that percentage is only expected to grow. The United Nations predicts that by 2030, 5 billion out of a total of 8 billon people on Earth will live in cities. In addition, most urban growth will occur in cities with populations of 500,000 to 5 million and the number of mega-cities (cities with populations of 10 million or more) in the developing world will surge. It is estimated that by 2030, the population of Africa's cities will be about 748 million, greater than the population of all of Europe (685 million).[15]

Many people will look at these statistics and claim that there is a bleak future for humanity. They will insist that we need to exact severe measures to control population growth including limiting births per couple or withholding medical treatment for persons over a certain age. We however, believe that while humans are the cause of many of the Earth's problems, we are also amazing problem solvers. We believe that if we are given the necessary information and enough time, we will solve and/or manage all of these issues. The concern is that we might wait too long to identify the threats and react.

The world population cannot continue to grow exponentially forever. It is simply not possible. Economist Kenneth Boulding wrote, "Anyone who believes exponential growth can go on forever in a finite world is either a madman or an economist."[16] At the same time, we agree with Professor Bruce Bridgeman that "it would be the ultimate tragedy if surging population numbers overwhelm our environment."[17] Once we realize the consequences of uncontrolled growth, we can sit down together and fashion a plan to deal with it. Once we implement an effective plan, the human race and our wildlife friends can look to a bright new future for many generations to come.

Depopulation

Before we leave the topic of world population growth, we must examine one more interesting twist. Many demographers and social scientists believe that the world population will reach 9 to 12 billion and then start to decline due to natural causes such as lower birth rates. Phillip Longman in his 2004 book, *The Empty Cradle,* explains that fertility rates are the key to what happens to the world's population. The replacement fertility rate, which is the number of children the average woman needs to bear for a population to sustain itself, is 2.1 Longman states that "global fertility rates have been declining for a long time. Today, they're half of what they were in 1972. Fifty-nine countries (accounting for 44 percent of the world's population) have fertility rates below replacement levels. The United Nations projects that by 2050, fertility rates in 75 percent of all countries will fall below replacement levels."[18] Even though fertility rates are expected to fall below replacement rates in many countries this does not mean that the world's population will suddenly stop increasing. It means that the rate of increase will slow, but absolute population numbers will continue to rise over time.

If these predictions are correct, it is likely that the population will rise over the next 50 years, perhaps beyond the carrying capacity of the Earth. Following this increase, the population will start a steep decline, but we do not know where it might stop.

Why is population growth so difficult to deal with?

Of all the problems facing our political, religious, business, and community leaders, finding ways to control population growth is by far the most difficult to solve. Why? It is not that we do not know where babies come from. The problem is that we are dealing with human beings.

If we were experiencing a sudden surge in the population of deer, for example, we would become concerned that there was not enough land and food to accommodate them. Action would have to be taken to slow their rate of growth. We deal with this issue all the time with wildlife. The solution most often used is to cull the herd to a number that is sustainable to the environment they live in. The solution is not so simple when we are dealing with people, but the problem should concern us even more because we humans consume far greater amounts of the Earth's resources than deer. At the same time, Americans believe strongly in the value of human life. Even discussing issues of human reproduction and death can be a dangerous line to walk.

What barriers must be overcome before we can begin dealing with this very serious developing world problem? The authors feel these can be reduced to three: (a) recognition of the problem, (b) dealing effectively with the human sex drive, and (c) reducing poverty through education. Let's first look at some of the reasons that stand in the way of effectively overcoming these three barriers.

First, one misconception most people have is that the world population is increasing at a slow but steady pace. This is not the case; the global population is expanding at an explosive rate. And, as with many of the problems identified in this book, the danger is not readily seen or felt by the people, especially in the United States, Europe, and Russia. We see a new child being born into a family as a good thing, something to celebrate. We realize that our communities are growing, but we have always associated growth as being a positive economic benefit to the community. The births come one at a time and it is difficult to see the larger picture of how adding more humans is a problem to our country or the world. This is always how we humans have lived and we have gotten along pretty well up to now. Why would we expect people to think that this is a problem?

Human population growth's real danger is that it will sneak up on us and once we recognize it to be a problem, the trend will be irreversible. The magnitude of the problem is different than humans have ever had to deal with in our history, so we have no experience to warn us or on which actions we can take to slow it down. These are new and uncharted waters. There is little wonder that we have a hard time coming to terms that we have a problem here; one that we must start addressing yesterday.

Secondly, the human sex drive is very strong and if people do not use contraceptives, they will have babies. Some of the methods that can be used to slow down the population expansions such as the use of contraceptives, the morning after pill, abortion, and decisions on terminating a life fly in the face of many people's religious beliefs both in this country and around the world. Any solution that challenges peoples' religious convictions is almost impossible for our political leaders to deal with in an effective manner.

Thirdly, we, for whatever reason, have not been listening to our scientists and demographers who have been trying to sound the alarm on pending population explosion for years. Perhaps this is due to the fact that the danger is considered too far in the future to be concerned about it, or perhaps it is because our political, religious, and business leaders have not identified it as a priority. Our guess is that there are two other reasons as well: (a) we are bombarded with so many predictions of pending disasters that we simply tune them out. The word "disaster" is over used and has lost much of its meaning; and (b) people feel it is a problem so huge that one person, even one government, cannot solve it and so they feel hopeless to even try.

Recommendations

The challenge for our political, religious, and business leaders will be to find a way to navigate the through these troubled waters of: (a) population explosion, (b) population contraction, (c) an ever increasing elderly population, and (d) reductions in both renewable and non-renewable resources. We need our scientists and our heads of State to inform us of pending problems. The scientists and demographers have been trying to sound the alarm on the consequences of run-away population growth but we have heard surprisingly little from our leaders. Prime Minister Tony Blair and President Vladimir Putin are notable exceptions to this general statement as both have spoken of the threat of the world's population explosion and the declining populations in their own countries.

The best possible answer is for us to begin addressing the impending problem. To do so, however, we must have leaders with the courage to educate all of us about the seriousness of the problems, explain that we must accept slowing the rate of world population growth, and accept limits on material consumption. These are not the kind of messages that people will want to hear and the leaders will face strong and hostile reactions to the sacrifices that they must call upon all of us to make. However, if we are to ensure that future generations enjoy the same standard of living these are the types of sacrifices we must make. If we start now, the changes required will be a lot less painful than if we choose to wait.

If we are able to convince our leaders of the seriousness of the issue what can we do to manage this problem? Two big actions are to work to reduce poverty around the world through education and introduce and/or increase family planning ideas in developing nations. Statistics show that reducing poverty and increasing education naturally slow population growth rates.

Education is the major key to reducing poverty and slowing the rate of population growth. There are approximately 860 million adults in the world that are illiterate.[19] It is not surprising that women make up two-thirds of that number. Over 50 million children around the world do not have access to formal education. These groups therefore have no way to fully comprehend the magnitude of this issue. The burden for educating these individuals will often fall onto the shoulders of developed nations. These countries have the resources to provide education, especially to those immigrants coming from developing nations. If we want these migrants to contribute to society we must work to provide them with the required education and training so that they can be productive citizens.

It was encouraging to see the proposals for the reduction of world illiteracy on the agenda of the Group of Eight's summit meeting for July 2006. These countries, consisting of Canada, France, Germany, Italy, Japan, Great Britain, Russia, and the U.S., are among the richest in the world and have come together with their resources to lead the rest of the world in the fight to eliminate illiteracy. Andrey Fursenko, the Russian Minister of Education and Science, offers his vision of the G8 nations forming a global educational program saying, "just imagine what the G8 nations can do now by putting our heads together. Perhaps we could create a global young educators program, sending our brightest young graduates to teach and develop literacy programs in the world's poorest nations. Perhaps we could bring together the best researchers in our universities to collaborate to develop solutions to ridding the world of illiteracy as we have childhood diseases such as smallpox and polio. Any number of approaches might work, but doing nothing is not an option. In Russia, we have a proverb that says 'Education is light, lack of it is darkness.' It is time to shine a light on illiteracy and vanquish the darkness. Now and for future generations."[20]

Family planning in developing nations is also important if we are to slow the rate of population growth. Education, urbanization, and contraceptive methods were vital in reducing growth in developed nations. Education, especially for women, is the best and strongest tool we have for bringing the world's birthrates down to a sustainable level. As women are educated and assume professional positions, they generally have fewer children, especially if they move into an urban environment. As countries become richer, their birth rates plummet. Since most of the population growth will be taking place in some of the world's poorest countries these people are unlikely to see substantial improvements in their economic standing. This brings us back to education on preventing unwanted births and supplying these countries with tools of prevention. These issues, nonetheless, can only be addressed effectively at the federal level and the U.S. needs to take a more active role in spreading this message along with Great Britain, Russia, and the United Nations. Another method to start reminding people about is adoption. In recent months many celebrities have been involved in high-profile adoption cases of children from Africa and Asia, reminding people that there are thousands of children already born that are in need of homes.

Conclusion

Controlling population growth will undoubtedly involve addressing unpopular and sometimes explosive topics. These issues will not get any easier with time and many people, including members of religious, environmental, and conservation groups are becoming deeply concerned about the consequences of doing nothing to slow this growth. It is time that our politicians, businesses, and community leaders get interested as well. One encouraging development in recent years has been the leadership of Bill and Melinda Gates, with help from Warren Buffet. Their foundation has spent millions to lead an effort to improve health and education in developing nations, as well as in the United States. Working together we can reduce and manage population growth in years to come.

CHAPTER TWO:
United States Population Growth

Question:
What will the United States population look like in forty years?

Answer:
There will be a lot more people; they will be older and more diverse

THINKING LOCALLY

Now that we've had the opportunity to think globally, let's spend some time focusing on the local issues. While population growth around the world will most definitely impact all of us here in the Unites States, most of us won't have much of a chance to affect what goes on outside of our borders. We will, however, have to make important decisions about how we want to respond to population growth in our own communities. As reported in the previous chapter, the majority of developed nations in the world are experiencing declining birth rates and these rates are likely to continue to decrease. The United States, however, is an exception to this norm. According to the U.S. Census Bureau, the United States has the highest population growth rate of any industrialized nation in the world and is currently growing by about 3.2 million people every year.[21] Our nation has added 100 million people since 1967. We are now the third-most populous country after China and India. In October of this year, the U.S. reached a population of 300 million and is predicted to exceed 419 million by the year 2050.[22]

How much growth will we experience?

The United States is, and will continue to be for the foreseeable future, the third largest nation in the world. We've come a long way from a population of 3.9 million, recorded in the first national census in 1790. In the most recent census in 2000, we had a population of over 280 million and our growth is not slowing. The Census Bureau is predicting a population of over 419 million by 2050. Table 2-1 depicts population projections for each decade between 2000 and 2050.[23]

Table 2-1: U.S. Projected Populations (2000 – 2050)

Year	Population (in millions)
2000	281.4
2010	308.9
2020	335.8
2030	363.6
2040	391.9
2050	419.9

This kind of growth becomes even more shocking when compared to past census data, dating back to 1790. Figure 2-2 illustrates this growth and indicates where we are currently.

Figure 2-2: Historical and Projected U.S. Population[24]

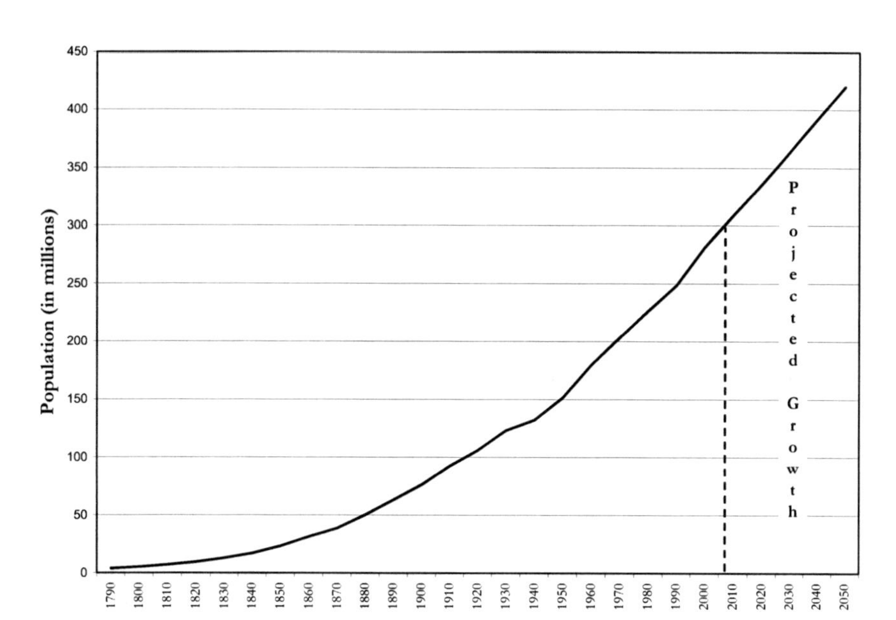

How will our population be structured?

At least for the next 45 or so years, the U.S. population is going to be both top and bottom heavy. The 2000 Census reported that over 29 percent of the total population was comprised of children age 18 and under; the elderly, age 65 and over, made up an additional 12 percent. This means that over 40 percent of all Americans fall into one of these two categories, and the number is going up. Table 2-3 shows a breakdown of the population by age.[25]

Figure 2-3: Population by age Group (2000 – 2050)[26]

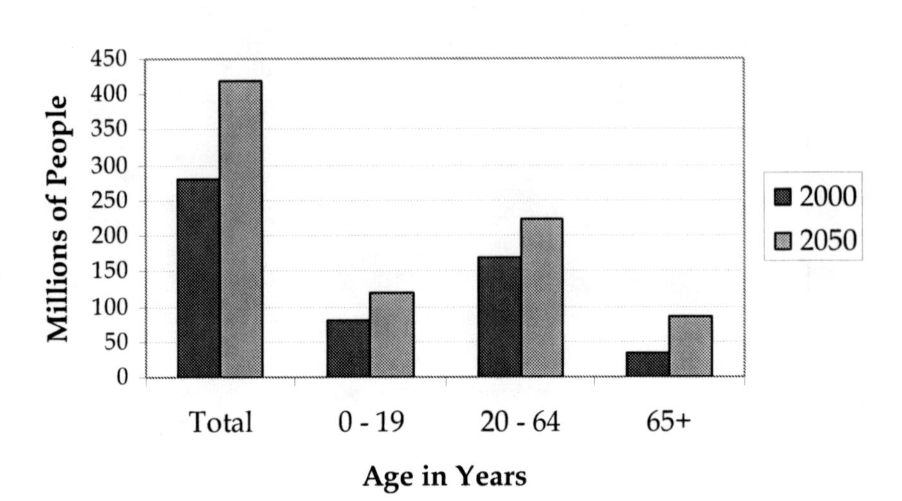

With such large numbers of old and young, communities are going to have to think about services differently. The elderly and children, for example, are more likely to need access to amenities such as additional day care and alternative forms of transportation. As their incomes often decrease with retirement, the elderly are also likely to need more opportunities for affordable housing and health care. These are only a few of the implications of planning for these particular age groups. Let's take a closer look at both groups, starting with the elderly.

Elderly Americans

According to a report published earlier this year by the Census Bureau, our nation's elderly population will "promise to redefine what it means to grow older in America." The elderly, defined as the population age 65 and older, is expected to double in size in just 25 years. This means that by the year 2030, almost 1 out of every 5 Americans will be elderly, a significant change to the nation as a whole. The report also discovered that the "age group 85 and older is now the fastest growing segment of the U.S. population."[27] With advances in health care, Americans are living longer than ever before. We need to ask ourselves whether we are building communities that will meet their future needs. Do we have the infrastructure and services in place that they will require? Will we be able to provide them with an adequate number of living options?

The two greatest challenges in providing for the elderly will be in providing retirement income and health care

Retirement:

The authors believe the greatest task for all levels of government, but especially the local governments will be dealing with the financial condition of the elderly. As we have discussed early in this chapter, there will be more elderly in the future and a greater percentage of the elderly will be females. Retirements can last for decades but funds from the retirements systems may not. Many elderly may simply out live their retirement funds.

Most companies have done away with defined benefit retirement plans in favor of 401(k) type plans. This may prove to be financially disastrous for many seniors as they reach retirement age. The defined benefit plans provided a fixed monthly amount the retiree would receive from the day of retirement until he/she died. Most plans provided a joint survivor's option that allowed the spouse to receive the retiree's monthly payments until they died. The defined benefit monthly payments combined with Social Security payments could be counted on for life.

Not so with 401(k) plans. This retirement plan guarantees the retiree only what is in the account at retirement. Under this type or retirement plan, the employee puts aside a certain amount of money each pay period, often matched by the employer, to be invested in stocks and bonds during the employee's working years. If the investments have done well, the retiree may live comfortably thought out his/her retirement years. Upon the death of the retiree, what remains in the investment account becomes part of the retiree's estate. However, if the investments have not done well or the retiree did not set aside enough money to invest, he/she may outlive the retirement funds.

The 401(k) plans are more risky for Americans than the defined benefits plans for two reasons: (a) Americans are not good savers. They seldom put aside the money needed for retirement. And, (b) because of its complexity retirement planning simply overwhelms many people and thus they don't do it.

The 401(k) is now the principal retirement plan for most Americans. It is much cheaper for employers to offer this plan than a defined benefit plan because there is no continuing financial obligation to the employee once he/she retires. Given that the 401(k) plan is the employee's retirement plan, one would think more attention would be paid to the amount needed for retirement and the investment decisions being made with the funds. Not so.

According to Fidelity Investments, more than one-third of corporate employees never join their company's 401(k) plan. If they do join, they do not save as they should and they pay less attention to how their 401(k) investments are doing than they pay to buying a new car. They are also quick to borrow from their retirement nest egg.

Health Care:

Even if Congress saves the Social Security benefits for the elderly many of them are going to out live their 401(k) funds. Once this happens, the elderly will have to depend on moving in with family members and on receiving all kinds of assistance ranging from health care to public transportation to recreation programs from the government. Health care will be the most troubling of all for families and governments to deal with. As we stated earlier, ageing brings on both physical and psychological changes. Depression, loneliness, boredom, decreased eyesight, hearing loss, weakened muscles, brittle bones and arthritis are common conditions. The elderly also face greater risks associated with heart disorders, strokes, and pneumonia, as well as the dreaded and scary Alzheimer's disease. There are currently 5 million people in the United States with Alzheimer's and this number is forecast to increase to 15 million by 2050. As medical technologies and advanced treatment extend our lives, new health care concerns arise for the elderly. One can only imagine the financial stress that caring for the elderly-- especially if they have exhausted their retirement benefits-- will bring to families and governments alike.

As more and more people age into the elderly group local governments will need to address the vital issue of their increased need for health care. Do we have enough hospitals and medical resources at their disposal? Are we providing them with ways to receive the medical attention that they need? Even if Congress is able to save the social security system, their increasing life spans will continue to make it difficult to provide the necessary resources. Local governments will be called upon to provide many services to these individuals. A higher percentage of the elderly may also decide to move in with family members, which means these families will need access to resources and information about their many options. Local health departments and human service agencies will most likely be the "first stop" for those seeking information and answers to their questions. Local programs such as Meals on Wheels, Homebound Services, and other counseling/support groups will be relied upon by local residents.

Actions that would help:

If Congress would place their members under the Social Security program and a 401(k) retirement plan instead of staying with their extremely generous retirement plan (which is a defined benefits plan) much more attention would be given to this very real and grave problem of the elderly out-living their retirement monies.

Universal health care for the young and the elderly will be an absolute must in a few years. It will come as the elderly population becomes a greater political force.

A Younger America

While the elderly population will continue to grow, by 2050 the percentage of the population 18 and under will be greater than the population 65 and over.[28] This too will challenge local governments, as they look for new ways to provide them with adequate health care, quality education, and eventually career opportunities. In addition, children 18 and under are more likely than any other age group to live below the poverty line.[29] These problems may require an increase in our social programs, targeted toward this age group. Both the elderly and children are constituency groups often forgotten in policy making, but whose numbers will continue to climb, as will the impacts of their growth. The elderly, who vote in large numbers, will be an increasingly powerful political force.

How much will the population's race/ethnicity change?

The United States will become a much more diverse nation by 2050 than it is today. The non-Hispanic white population will decrease from 69 percent of the population in 2000 to 51 percent in 2050. The nation's Hispanic and Asian population will continue to grow at a much faster rate than the non-Hispanic white race and increase their percentages in the total population; Hispanics will grow from 12 percent to 22 percent and Asians will grow from 4 percent to 11 percent by 2050. The African American population will also grow as a percentage of the population between 2000 and 2050 from 12 percent to 15 percent. These population changes in race/ethnicity are shown in the figures below:

Figure 2-4: U.S. Population by Race/Ethnicity (2000)
Total Population – 281.4 million

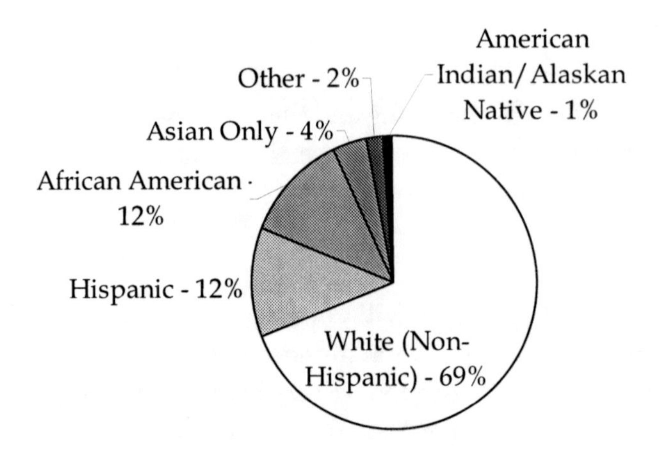

Figure 2-5: U.S. Population by Race/Ethnicity (2050)
Total Population – 419 million

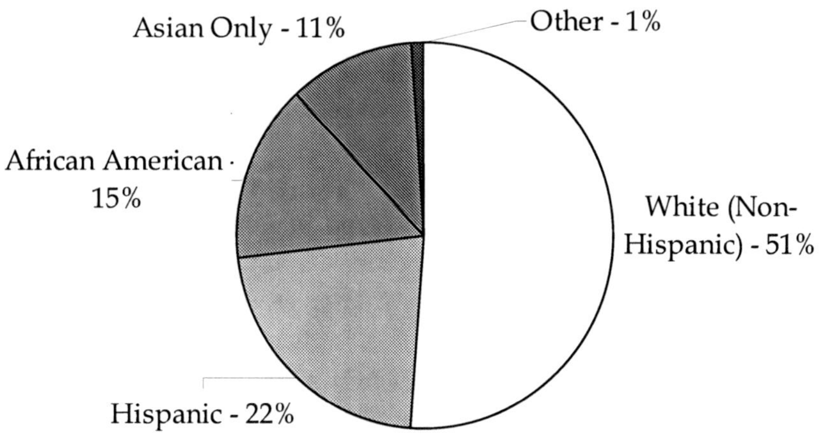

Asian Only - 11%
Other - 1%
African American - 15%
White (Non-Hispanic) - 51%
Hispanic - 22%

WHERE WILL THIS GROWTH OCCUR?

While the U.S. will see significant growth in the near future, that growth is going to be concentrated primarily in the southern and western portions of the country. From 1970 to 2000, the percentage of Americans living in the south and west grew from 48 to 58 percent.[30] The west alone surpassed the population of the northeastern United States in 2000 and is expected to surpass that of the Midwest in less than 25 years. At the same time, more people than ever before are moving to urbanized areas of the country. Today, 50 percent of Americans live in suburbia and another 30 percent in our central cities.

Regardless of regional differences, individual states are going to be affected disproportionately by this growth. Seven states, including Texas, Florida, and Georgia, are expected to have annual growth rates of over 2 percent, which means they will double their populations in less than 35 years.[31] Not surprisingly, California is expected to lead the nation as the fastest growing state through the year 2025. The Census Bureau has predicted that the state will increase its population by close to 18 million between 1995 and 2020, which means that in less than 20 years the state will single-handedly house

15 percent of the nation's population.[32] Given this uneven growth, some communities will feel the effects of population increases sooner and to a greater extent than others.

What is driving this growth?

The biggest driver of this growth is immigration, both legal and illegal. Immigration has been a hot topic of conversation recently. The debate surrounding illegal, and legal, immigration has reached the cover of every major newspaper and magazine in America and has led to proposals for new legislation in Congress. These proposals have the potential to substantially change immigration laws for the first time in decades. But what role does immigration really play in America? Ignoring many of its social implications for the moment, how is it really going to impact our population in the future?

Population growth in the U.S. since 1990 has been driven primarily by immigration. It has been estimated that between 40 and 60 percent of the growth in this country over the past two decades can be accounted for by immigrants and their children. While the fertility rate in the U.S. is the highest in the industrialized world at 2.09, it is due primarily to the higher reproduction rates of immigrants.[33] The American fertility rate has actually been falling for the past 200 years. In 1800, the white fertility rate was 7.4, which is over three and half times the 2.07 of today. White fertility rates continue to decrease and have already fallen below the sustainable rate of 2.1 previously mentioned.

To off-set the falling white fertility rates, the number of immigrants entering the county has been increasing. Once they arrive they also tend to have higher fertility rates. While the non-Hispanic, white population grew by only 0.3 percent between 2004 and 2005, the "Latino and Asian ethnic groups… [grew] at rates of 3.3 percent and 3 percent respectively."[34] Without immigration our population would see a growth rate of only about 15 percent, which would put our population at just over 325 million by 2050 – a great deal less than the 419 million projected.[35]

Table 2-6: Immigration and U.S. Population Growth[36]

Year	If immigration continues at current rates, population will be...	Percent change from population in 2000 (%)	Without immigration, population would be...	Percent change from population in 2000 (%)
2005	295,870,000	5	288,580,000	3
2010	310,610,000	10	295,770,000	5
2015	325,400,000	16	302,880,000	8
2020	339,630,000	21	309,320,000	10
2025	353,120,000	26	314,530,000	12
2030	366,010,000	30	318,350,000	13
2035	378,160,000	34	320,950,000	14
2040	390,160,000	39	322,670,000	15
2045	401,830,000	43	324,000,000	15
2050	419,900,000	48	325,580,000	15

Since immigration is predicted to play a major role in how our population will grow between now and 2050, policies to direct and control the flow of immigrants into the United States are critical. The upside is that we can control part of our destiny by our immigration policies.

A Short History of Immigration

During the Industrial Revolution (1800s to early 1900s) the United States needed large numbers of skilled and unskilled workers to fill its industrial jobs. There was no such thing as an illegal immigrant as everyone, with a few exceptions, was welcome. Around 1919 to 1920, however, concern developed over the number of immigrants that were in the country and Congress, reacting to that concern, passed laws restricting the number of newcomers that would be allowed to enter the country each year. It is interesting to note, that the citizens who lobbied Congress to pass the new immigration laws argued that immigrants stole jobs from current Americans, were ignorant, criminals, and showed no interest in becoming citizens.

Illegal immigration started with the passage of this legislation. Immigration did not stop, nonetheless, as individuals continued to enter the country without permission. The number of illegal immigrants became large enough that Congress again felt that they needed to take action. With the passage of the Registry Act of 1929, Congress established quotas as to the number of people who could immigrate to this country from various nations. Immigrants who had arrived before 1921, however, were allowed to become citizens by paying a $20 fee.

From 1935 through the late 1950s, thousands of Europeans who were unlawfully in the U.S. were allowed to temporarily travel to Canada and re-enter the U.S. legally as permanent citizens. At this time there were no quotas for Canada, Mexico, or other western hemisphere countries. Then, in 1965, Congress placed quotes on all countries, including those in the west.

History repeated itself in 1986 when citizens once again became upset over the millions of illegal immigrants living in the U.S. In response, Congress passed legislation creating an amnesty program that would allow 4 million illegal immigrants to become legal citizens. This legislation was to end the illegal immigration problem once and for all. It called for: (a) increased control of our borders, (b) tight controls over those entering the country, (c) tougher enforcement for those who allowed their visas to expire, and (d) penalties on employers who hired illegal citizens. The problem was no one enforced these regulations.

We learn from history....but not much!

Now, in 2007, to quote Yogi Berra, "it's deja vu all over again." Today, there are a reported 12 million illegal immigrants in the country causing a heated and emotional debate as to what to do about them. President Bush has a plan to:

- Provide the illegal immigrants with a pathway to citizenship;
- Establish a guest-worker program;
- Secure the borders by building a 700 mile fence along the U.S. and Mexico boarder and place 6,000 National

Guardsmen and new border patrol Officers to stop illegal crossings; and
- Penalize employers who hire illegal immigrants.

Congress has not, at this time, passed the President's immigration proposal and does not seem inclined to do so.

There does not appear to be any disagreement that the illegal immigration problem needs to solved and solved soon. However, there is no consensus as to how to solve the problem. One side argues that because of our aging population and the retirement of the baby boomers we need immigrants to fill jobs and help pay the costs of social security for retired workers. The other side counters that the estimated 12 million immigrants that have come across our borders illegally actually put a strain on our economy by using medical and educational services, paying fewer taxes, and "taking away" jobs from Americans. Interestingly, many illegal immigrants in this country actually file tax returns and pay regular taxes on their income.

Chain Migration

A provision of the United States' immigration law, referred to as the "chain migration" provision, allows not only a new citizen's immediate family to become citizens, but also the grandparents, adult sons and daughters, and other relatives. The result is that one new citizen can translate into a dozen in a real hurry. There has been little to no debate on how many immigrants, legal or illegal, the country can actually accommodate. This is one of the most important issues Congress and the President will be called on to make for the country. The United States has long equated growth with prosperity, but is there a tipping point where growth makes us less prosperous? Can the country accommodate a population of 300 million and still maintain its high standard of living? How about 419 million?

To get a better idea of just how large an increase of 119 million people would be for the U.S., consider that the entire population of Mexico is 107 million. This means that in the next 43 years, according to U.S.

Census Bureau estimates, our population will increase more than the entire population of Mexico today. That is a lot of newcomers!

Recommendations

It would be nice if we had room for everyone in the world who wanted to come to the United States -- but we do not. The first decision that we as a nation must make is how many people we can accommodate on a sustainable basis. Once that decision is made, we can determine the total number of immigrants we want/need. In the future, immigration policies that are only geared towards providing a constant supply of low-paid, unskilled workers will not serve the country well. We need scientists, engineers, doctors, and other highly skilled workers as much as we need unskilled ones.

From a strictly personal view, the authors do not feel that illegal immigrants should be allowed to stay in this country under amnesty programs. This is a very hard decision as so many of these illegal immigrants are good, hard working people who came to this country in search of a better life for their families. However, we do not believe that we should reward people who have broken our laws and "jumped" in the line ahead of those who have played by the rules. That being said, we do feel that Congress must create an easier to administer guest-worker program. Congress should make it the responsibility of the businesses hiring the workers to stop paychecks when workers' permits expire and to notify the government when it is time for these guests to return home. Harsh penalties on employers who hire illegal immigrants must be enacted and enforced. Since a majority of our growth between now and 2050 will come from immigration, we can control much of our population growth if we choose to.

While immigration decisions are made at the federal level, there are many steps that communities can, and should, take to begin planning for growth. Everyone can begin by just taking a closer look at population projections for their cities/towns. Demographic studies can be fairly easy, and quite inexpensive, to administer using cohort-component modeling. They can also be extremely valuable in understanding how the population structure of individual communities will change. We can determine what areas of our communities will grow the most, which age groups, and even which ethnic/racial groups. Once we have a clear picture of the size and make-up of our future population, we can begin thinking about the implications on land use and service requirements.

Conclusion

These projected changes in population raise questions for local government officials as they will forever change the way in which we operate in communities across America. We will have to ask ourselves many of the same questions that nations around the world will have to ask. Do we have enough land to accommodate approximately 120 million more Americans? Can we feed them? How will we house these new residents? How will governments provide them with services? How will the U.S. accommodate another 80 million additional cars on our roads? All of these questions are important ones that must answered by local governments as they are the primary providers of public programs and services. The key is to ask ourselves if we can learn to adapt to and accommodate these changes, while still maintaining our standard of living.

CHAPTER THREE:

Global Warming

Question:
Is global warming real, or a great hoax?

Answer:
It is for real. The Earth is warming! The armadillos and the Eskimos know it.

IS THE EARTH IN DANGER FROM GLOBAL WARMING?

The Earth itself is not in any real danger. There have been numerous climate changes in the Earth's history. We are aware of six ice ages (four within the last 450,000 years) with multiple warm periods in between. There will doubtless be future periods of climate change, with or without the influence of human beings, which the Earth will survive. It is the Earth's populations, therefore, that will be adversely affected by climate change including: humans, animals, aquatic life, and plants.

BACKGROUND

Are we experiencing global warming as many atmospheric experts warn? Or is this as Bill Gray, the noted scholar and professor emeritus at Colorado State University believes, "one of the greatest hoaxes ever perpetrated on the American people"? At this point is seems impossible for governmental decision makers and the public to know for sure. On both sides of the issue there are many knowledgeable, well-respected scientists who espouse opposite viewpoints. Those who agree with Bill Gray point out that many of the predictions about global warming do not come from actual temperature measurements and greenhouse physics; rather they come from manmade computer models relying on numerous assumptions and guesswork. They claim that these models have not been validated against historical temperature records and that the Earth will begin cooling again in the next five to ten years.[37]

Those on the other side of the issue, however, are convinced that the Earth is warming and that there are serious consequences if we choose not to act. Dr. Richard S. Lindzen, Professor of Meteorology at the Massachusetts Institute of Technology, believes that greenhouse gas emissions are in fact contributing to global warming. At the same time, he agrees with Gray that relying on computer forecasting models is unwise. He writes that "many of the most alarming studies rely on long-range predictions using inherently untrustworthy climate models, similar to those that cannot accurately forecast the weather a week from now."[38] Further complicating the matter is a recent discovery by scientists claiming that years ago the "North Pole was a lot like Miami, with an average temperature of 74 degrees [Fahrenheit], with alligator ancestors and palm trees."[39]

Jim Hansen, Director of the NASA Goddard Institute for Space Studies, points out that while humans have paid little attention to the issue until recently, scientists have observed it through plants and animals for many years. In order to survive, many species of plants and animals have been migrating towards the Earth's poles because of changes in their climate zones. If species are unable to adapt to changes, they must migrate for their basic survival. In *The Threat to the Planet*, Hansen writes the following:

> "Studies of more than one thousand species of plants, animals, and insects, including butterfly ranges charted by members of the public, found an average migration rate toward the North

> and South Poles of about four miles per decade in the second half of the twentieth century."[40]

Hansen also points out that while this migration may seem significant, it is not occurring fast enough. Species like the armadillo, unable to contend with temperature increases, are traveling at up to 35 miles per decade in this effort to survive. The odds are that the armadillo will not be able to migrate fast enough to make it. Hansen worries that up to 50 percent of all species could become extinct.

The truth is that no one has the "answers" to global warming or climate change questions. No one knows for sure what the future holds or how this issue will play out. We do not know whether global warming is a result of human actions, or a natural cycle of the Earth. Twenty years ago most of us didn't even know anything about it, but today it has risen to the mainstream consciousness of the public. As a result, everyone is searching for answers and strategies to combat predicted changes. We believe the evidence indicates that global warming is a reality we will have to deal with. For those that do not agree we pose a question. Are any of the strategies recommended to combat global warming damaging to your city or town? Will green buildings or alternative energy sources be detrimental to your community? Will investing in public transportation systems negatively impact the public? These solutions may cost us more today, but will we reap greater benefits in years to come?

If global warming is in fact a reality, local governments need to be asking themselves (1) how global warming will affect the quality of life in their community and (2) how they need to respond to the situation. As we do not know the extent to which the Earth will warm these questions may be difficult to answer. Estimates on temperature increases range from 2.7 to 11 degrees Fahrenheit, which can make a tremendous difference for humans and other forms of life. Higher temperatures may also equate to melting ice caps, rising ocean levels, and flooding for many low-lying areas of land. The last time that the Earth was five degrees warmer, for example, the oceans were 80 feet deeper on average. If this occurred again, Florida, the Netherlands, Bangladesh, and many cities on the coasts would be underwater and millions of people would be displaced from their homes. This would also lead to a decrease in the food production potential and the endangerment of wildlife species around the globe. The effects are seemingly endless and may require some serious adaptation on our part.

The Greenhouse Effect: A Primer

Whether global warming is "man-made" or part of nature's cycle, the primary reason for the rise in global temperatures is thought to be a result of steep increases in greenhouse gases in the atmosphere (primarily carbon dioxide and methane). Our use of fossil fuels, in numerous capacities, has certainly added to the rise in carbon dioxide levels in the atmosphere. But what is the greenhouse effect and how does it impact us?

The greenhouse effect is a natural process, essential to human and animal life on Earth. Energy from the sun warms the Earth as it absorbs the heat provided. The Earth in turn radiates some of that heat back into space in the form of infrared radiation. Approximately one percent of the Earth's atmosphere is composed of greenhouse gases, primarily in the form of water vapor, carbon dioxide, methane, and nitrous oxide. Together, these gases reflect enough of the heat back to Earth to maintain the average temperature of the atmosphere (around 60 degrees Fahrenheit). As we emit more and more greenhouse gases back into the atmosphere, greater amounts of the gases are captured and more heat is therefore returned to Earth. This process is critical to maintaining the Earth's temperature. Without it, the Earth would be a cold, uninhabitable place.[41]

The key, however, is balance. We need to have a particular level of greenhouse gases to keep the Earth warm enough, but too much can lead to significant change in our climate. Records taken from ice cores indicate that whenever carbon dioxide levels have increased in the Earth's past, so have temperatures. Current concentrations of carbon dioxide in the atmosphere hover around 380 ppm (parts per million), which is the highest concentration known to have ever existed.[42] In addition, total global carbon dioxide emissions from the burning of coal, natural gas, petroleum, and wood have grown from a few hundred million tons of carbon dioxide in 1906 to seven billion tons today, or more than one ton per person on the planet. Americans, as you can imagine, are by far the greatest polluters, averaging around six tons of carbon dioxide emissions per year. Methane levels in the atmosphere have also been on the rise. Within the last 100 years, methane levels have doubled and we now emit about 540 millions tons each year.[43]

Many may assume that rising temperatures only affect land masses, but ocean temperatures are also increasing. Increasing levels of carbon dioxide in the atmosphere also causes greater amounts to dissolve into our oceans. Dissolving carbon dioxide in water increases the acidity of the oceans and has been linked to the destruction of coral reefs.

While temperatures in general are increasing, nighttime temperatures are actually rising more quickly than daytime temperatures. This fact is primarily due to urbanization resulting in fewer trees and greater amounts of impervious surfaces, such as concrete and asphalt. These pavements hold in higher quantities of heat during the day and as a result do not cool quite as much during the evening. Our days are therefore more likely to begin at higher temperatures. Consequently, demands for water and energy for air conditioning are greater.

How do we impact the greenhouse effect?

As we mentioned above, a certain level of greenhouse gases in the atmosphere is natural. Humans however, produce additional amounts of these gases as a result of our daily living. Examples include some of the following:

- Agriculture and Landfills – Agricultural activities and our landfills produce large amounts of methane gas, which occurs when bacteria decomposes organic matter. Examples of this include the decomposition of animal manure on farms and the disposal of urban waste into landfills. It has been estimated that 25 percent of the methane gas in the atmosphere is a result of human activity. According to the Environmental Protection Agency (EPA), methane is more than 20 times as effective at trapping heat in the atmosphere as carbon dioxide.

- Deforestation – Trees and other plants naturally remove carbon dioxide from the air. The destruction of many large forests has reduced the ability of plant life to clean the air. In addition, as forests are cleared, a large percentage of the trees are burned or left on the ground to decompose. This burning and decomposing allows additional greenhouse gases to escape into the atmosphere.
- Fossil Fuels – The burning of fossil fuels (i.e. coal, natural gas, and oil) accounts for approximately 75 percent of our carbon dioxide emissions. We burn fossil fuels to produce the majority of the energy for electricity, heat for our buildings, and power for our cars.[44]

The United States and China emit, by far, the greatest amounts of greenhouse gases. Table 3-1 below illustrates carbon dioxide outputs in 2002 by country:

Table 3-1: Carbon Dioxide Emissions by Country (2002)[45]

Source Country	Metric Tons of Carbon Dioxide Emissions
United States	1,576,000,000
China	1,033,000,000
Russia	419,000,000
Japan	331,000,000
India	302,000,000
Germany	236,000,000
United Kingdom	148,000,000
Canada	141,000,000
South Korea	136,000,000
Italy	123,000,000
Mexico	108,000,000

Why is global warming so difficult to handle?

The amount of material written on the subject of global warming is incredible. Scientists around the world, and from both sides of the issue, have been trying to get the attention of the public for years. Why has it taken us so long to take notice?

The first reason is that the issue is a complicated one, and there is not a consensus as to the prognosis or the solution. As theories and ideas are shared, holes are poked in them and new ideas are born. Like so many things in the world, certainty is nonexistent. In addition, the changes will probably happen very slowly. This means that we won't be able to actually feel a difference in temperature or see the developing consequences. If, for example, we experience a five degree increase in temperature, the sea level will rise approximately 80 feet. This will equate, however, to only 20 feet every 100 years.[46] At this pace, we can hardly expect the average individual to detect change or become concerned over the matter.

The second reason is that we have never before faced a threat such as this one. We have no history or experience to fall back on. At the same time, we tend to focus on the short-term, not the distant future. If people do not see an immediate threat they will not want to change their lifestyle. Larry King once stated, "no one cares about fifty years from now," and it's true. But we need to begin planning for the future. Many fear that global warming has the potential to cause irrevocable changes that we will not be able to reverse. "Global warming may, in effect, be fueling global warming. We could be on the verge of a tipping point at which climate change shifts from a gradual process that can be forecasted by computer models to one that is sudden, violent, and chaotic" says Worldwatch President Chris Flavin.[47]

Lastly, global warming has become not only a scientific problem, but a political one as well. Even if scientists develop ways to offset the effects of global warming, it will still require us to reduce our dependence on fossil fuels. In the end, this will require the action of our elected officials. There are numerous large, politically well-connected companies that benefit from lax emissions standards. And these are the companies that make significant campaign contributions to many politicians at all levels of government! They hire experts to say that global warming is a hoax and at the very least create doubt in our minds as to whether it is real or not. Then, instead of setting strict, required standards, we ask for "voluntary compliance" or pass legislation that goes into effect many years down the road.

In all fairness, global warming is a difficult issue to deal with, but we must do something. We need to admit that there is, or could be, a serious problem and take steps to ensure our future.

What can we do about global warming?

All nations must recognize that global warming is occurring and concerted actions are needed now to lessen the negative impacts. Many European countries, South Africa, Australia, and New Zealand, for example, are already taking the lead in this effort, while the United States continues to lag far behind. In Great Britain, Gordon Brown, Chancellor of the Exchequer and most likely the next Prime Minister, has urged the establishment of a global emissions trading system to cut carbon emissions. He is also calling on other nations to join Britain in establishing a new $20 billion facility to diversify energy supplies for developing nations.[49] In the U.S., however, federal level action is not occurring. President Bush does not support governmental action preferring instead voluntary efforts by industries.

Voluntary actions

Until the United States becomes a major player in combating global warming, the problem will not be solved. The present administration prefers to back voluntary actions by well-intended corporate leaders to reduce carbon emissions. In the authors' opinion this is wishful thinking. *The Week,* reporting on a June 2006 article in *The Washington Post,* stated that "of the 74 companies that joined the Bush administration's voluntary 'Climate Leaders Program', half have still not even set targets for reducing their greenhouse gas emissions."[50]

Voluntary action simply will not get the job done. It is Economics 101. For companies, the bottom line cannot be ignored for long. Regardless of how much they want to take actions that are good for the environment, if it adds costs that make companies uncompetitive they feel that they cannot afford to do so. Companies can not afford to put themselves in a non-competitive position. Thus, if their competitor is not voluntarily adding cost to help the environment, they will have a hard time doing so and the system falls apart. However, if government requires the reduction of greenhouse gasses or some other environmental protection measure that all industries must follow, than the playing field is more level.

Smoking bands in restaurants are a good example of how government has to place all establishments under the same rules to be effective. Restaurant owners feel that they cannot afford to have a smoke-free environment if other restaurants allow smoking. Thus they will not voluntarily do so. Once, all restaurants are required to have their establishment smoke-free everyone is again in a position of equality. It will take governmental action and laws to even the playing field so that companies can afford to make the expenditures necessary for them to reduce their greenhouse emissions.

The sad truth is that unless the United States takes a lead on the world stage and solicits the support of China and India to do likewise, there will be no substantial reduction in greenhouse gas emissions and the world will continue on its warming ways. The United States and China must work together on making deep cuts in their greenhouse gas emissions. It is up to the U.S. to lead the way. We are the largest polluter and without a strong commitment from us to cut these gasses, we have no credibility in asking or trying to get the world community to pressure the Chinese into taking action. The U.S. cannot ask other countries to adopt economic policies that promote "sustainable development," without adopting similar policies. Thus far, however, President Bush has argued that the cost of mitigating global warming is too large to be justified.

The Eskimos (Inuits) who live in the Arctic, however, would disagree. They are claiming that temperatures have risen so much that they now need air-conditioning. They have gone as far as filing a complaint with the Inter-American Commission on Human Rights claiming that the United States has contributed so much to global warming that it should be considered a violation of their human rights.[51]

Is the United States willing to lead the effort to fight global warming by reducing carbon dioxide emissions? The answer to date has been no. However, at the 2007 G8 Summit, President Bush and the other seven nation's leaders agreed to work together to reduce greenhouse gases. The President also pledged to take a leadership roll in this effort. This could be a very positive step in slowing global warming.

There are five categories of actions that nations can take to slow the rate of global warming:

- Reduction of energy use (conservation)
- Shifting from fossil fuels to alternative energy sources
- Carbon capture and storage
- Carbon sequestration
- Planetary engineering to cool the earth. [52]

Truth and Action

What is needed is for the United States to lead the world effort and the President, with Congress' support, to lead at home. In order to lead at home, the people need to know the truth and they need to be asked to sacrifice. We can not avoid sacrifice; however, if we take action now the sacrifices will be much less disruptive than if we leave it for our children or grandchildren to shoulder the burden.

The authors firmly believe that the American people will follow the President and meet the challenges if they believe that the situation is, or will become, grave without a drastic reduction of greenhouse emissions. Recent polls by the Yale Center for Environmental Law and Policy and the New York Times/CBS©[53] clearly confirm this. The people are way ahead of the President and Congress on this issue. They are ready to join in fashioning solutions for combating global warming. They want some straight talk and some strong leadership.

Once we know how the government plans on proceeding and on what time frame goals much be met, businesses, state and local governments, non-profit organizations, religious groups and citizens can all play a supportive role.

Carbon cap and trade or pollution tax?

There are two ways that the President and Congress could go about reducing greenhouse emissions: (a) levy a pollution tax, a tax on each ton of carbon dioxide emitted into the air, or (b) install a cap and trade regulatory system. The latter approach sets a limit for overall emissions with each individual polluter getting a permit setting their limits. Then the companies that are able to reduce their emissions below their permit limits could sell a portion of their "permit to pollute" to another company.

Legislation has been introduced in Congress to provide for a carbon cap and trade system. While this approach is the easiest to pass, it is also the less effective of the two approaches. One also has to be a bit suspicious when the largest polluters favor this approach.

A carbon cap and trade program simply allows everyone to continue doing what is being done now. Companies will buy the right to pollute and continue to do so. The Environmental Protection Agency recently decided to allow power plants that exceed mercury pollution caps to buy "credits" from plants emitting less pollution. This is a big problem as mercury is too dangerous to humans, especially pregnant women and young children, to allow any buying or trading which allows a plant to continue dumping unsafe loads of mercury into the air.

In addition, a cap and trade system would be complicated and hard to regulate. Unless the government sets up an enforcement body to keep everyone honest, the system would soon breakdown. With more enforcement comes more cost to the tax payers. The result is that we would have put in place an expensive program which at best will be slow, cumbersome and subject to manipulation; at worst, the program will be completely ineffective in reducing greenhouse gasses. The investigative report that follows outlines the difficulty of administering a cap and trade program that is both fair and effective.

In April 2007, an investigative story in the *Financial Times*© about carbon trading found that some organizations were paying for emissions reductions that did not take place while other organizations are making big profits from carbon trading. Specifically, the *Financial Times* reported:

- Widespread instances of people and organizations buying worthless credits that do not yield any reductions in carbon emissions.
- Industrial companies profiting from doing very little—or from gaining carbon credits on the basis of efficiency gains from which they have already benefited substantially.
- Brokers providing services of questionable or no value.
- A shortage of verification, making it difficult for buyers to assess the true value of carbon credits.[54]

So why a pollution (carbon) tax?

There are many good reasons for a pollution tax over a carbon cap and trade system. These include:

- It can be adopted in short order. There are no lengthy phase-in periods. No "politics of delay."
- People and industries respond quickly to economic incentives or disincentives.
- Congress does not have to design, monitor and enforce a cap and trade system.
- Congress does not have to set and enforce higher fuel efficiency standards for cars and trucks.
- The tax revenues can be used to financially assist citizens and/or industries switch to cleaner burning fuels.
- The tax will create the incentives to develop cleaner energy sources now rather than waiting 10 years for a governmental mandate.

- It would start immediately to reduce the country's dependence on foreign oil.
- If Congress will not adopt the tax, states can.

A pollution tax is much easier to regulate and control than a carbon cap and trade program and much less expensive for the taxpayer. Congress doesn't have to go through all the hassle and endure all the pressures from the auto and oil industry lobby because it doesn't have to pass higher mileage standards for cars and trucks; the consumer and the market place will demand more efficient vehicles.

Enacting a pollution tax in which everyone pays for the carbon dioxide they release into the atmosphere would create an incentive for people to look for solutions. It would set the innovative juices flowing and encourage long term investments in energy savings, carbon clean-up and new technologies. We would likely see technologies developed that we have not yet even imagined.

Because the pollution tax would create an immediate incentive to develop new technologies, new businesses would be created and the U.S. business community could take the lead in developing the "green businesses" of tomorrow that the world will need. Congress, by not passing strong environmental laws and providing strong disincentives for polluting, has encouraged U.S. businesses to "sit on their hands" which allowed European and Asian countries to jump ahead in this important developing industry. We need look no father than the auto industry to see that the federal government's failure to impose increased mileage standards allowed the Japanese to get a six-year jump on Ford©, Chrysler© and GM© in developing the hybrid automobile.

Politicians are afraid of anything that sounds like a tax. There are times when this fear of taxation causes Congress to hide the increase from the public. They do so by letting companies raise their prices that we all pay. However, the citizens know that cutting greenhouse gases is going to be costly. They know that they will be the ones to pay this cost either through higher taxes or higher costs from industries. Given the true facts, the citizens will realize that a carbon tax has a big advantage to the taxpayer and the environment over a cap and trade system. If the pollution tax is used to help citizens replace heating/cooling systems, purchase more fuel efficient vehicles or support research of pollution reduction, citizens will reap the benefits. However, with a carbon cap and trade system, the money all stays within the industries that are buying and selling the pollution permits. Citizens end up paying the costs but receive no benefits beyond carbon dioxide reductions. The companies that sell their pollution permits are the ones who make out financially.

According to the U.S. Energy Department, there are as many as 150 new coal-fired power plants that are scheduled to be built between now and 2030. Each new plant, if built by today's pollution standards, is bad public policy, bad global policy and bad economic policy. A carbon cap and trade system will not keep these plants from coming on line nor provide companies with the economic incentive to build clearer, less polluting facilities. Building a less polluting plant requires higher construction costs which companies are reluctant to pay for out of fear of becoming less competitive. A pollution tax on each ton of carbon dioxide emitted into the air would cause every polluter to pay the same amount. This would give the advantage to the less polluting plants thereby changing the economic formula and resulting in less polluting plants.

Three other things the President and Congress could do to combat global warming is to: (a) continue to fund basic research on ways to control or adapt to global warming, (b) develop a transportation system less dependent on the automobile by giving more transportation options to the people and (c) lead by example by putting the federal agencies on a carbon diet.

Research

The federal government needs to fund basic research into:

- Developing ways to conserve the limited supply of fossil fuel within the control of the U.S.

- Developing alternatives to fossil fuels to power vehicles and heat and cool homes, commercial and industrial buildings.
- Developing ways to use coal by eliminating or capturing and storing its carbon emissions
- Developing ways for humans, animals and plant life to adapt to a warming world.
- Developing new planetary technologies or engineering to help cool Earth.

Engineering a cooler planet

Some of the far-out ideas engineers are working on to help cool the planet include: carbon capture and storage, launching solar umbrellas into space to shade the planet, building artificial trees, and dumping volcanic dust into the atmosphere.[55]

Carbon capture

The federal government is now funding research in carbon capture, storage and sequestration. This is important research and should be continued; these approaches hold promise in reducing the carbon dioxide being released into the atmosphere. Coal is too important a fuel source not to do all we can to find ways to either burn it cleaner or to lessen the carbon dioxide that burning it emits into the atmosphere. Until new technology is available, all new coal fired plants should be placed on hold. The federal and state governments need to place a building and permitting moratorium on all such plants.

Transportation

The automobile is a wonderfully convenient from of transportation. Hopefully, the automobile can remain the backbone of our transportation system forever. However, to protect the automobile we must develop other transportation options. Why? The reason is that our roadways are congested today and given that we are expecting 119 million more people in the U.S. by 2050, we simply cannot build the highways and other infrastructure needed to serve this population. If our present trend in automobile ownership in the U.S. continues into the future that will mean an additional 80 million autos and small trucks on our highways. This does not include the additional large truck traffic needed to haul materials and goods across the nation to supply this new population. The plain truth is that we cannot build our way out of this situation by more and more highways. To protect the rights of all of us to drive our cars, we must invest in high speed trains and light rail systems to move more and more people.

The federal government, in partnership with the states, need to put the same vision, energy and funds into developing a high speed train system that spans the country as they did in building the interstate highway system.

States and cities, with federal financial assistance, need to build better public transit systems (bus, light and commuter rail) rather than more streets and highways. Urban populations are forecast to become increasingly elderly. The elderly will need reliable public transportation and they are living many years beyond the age they should be driving.

Lead by example: The symbolism would be huge

Mr. President, put the solar panels back on the roof of the White House! President Carter had solar panels installed on the White House and President Reagan removed them; putting them back into operation would be a wonderful visual symbol of the federal government's commitment to using alternative forms of energy. Let's make the White House the flagship in our fight to curb global warming.

Recommendations

The federal government has by far the greatest opportunity to take meaningful and effective actions to slow global warming. However, state and local governments also have important roles to play. State and local governments not only can set good examples by their actions, but they also have the power to reward local companies for their efforts as well. They can create programs similar to the federal programs, but obviously on a smaller scale. They can offer tax benefits, for example, to companies who reduce their emissions or energy requirements. They can assist companies in getting their employees to use public transportation or live in the community in which they work. The opportunities are endless if we are willing to form these partnerships and govern creatively.

Not all municipal efforts need to be so complicated. There are many simple, low-cost options for local governments that could be considered:

- Planting trees and passing ordinances that limit the removal of trees for certain types of development. Every city could strive to become "a city within a park." This is a great area to involve garden clubs, civic clubs, boy and girl scouts, school groups, university students, and a host of others all coming together to help the environment and make their city a more attractive and cooler place to live.

- Buy, and encourage gifts of, as much park land and/or open space as you can afford. Team up with land conservation groups to plan and maximize open space and park land. This is also a great way to encourage higher density development.

- Make certain that all municipal and state-owned buildings are as energy efficient as is financially feasible. Simple changes such as changing lighting systems or installing a few solar panels to provide for light, heating, and cooling can make a significant impact. In the long run, many of these improvements are also cost-effective. These actions prove leadership and can improve the credibility for elected officials when they ask citizens and corporations to make similar choices.

- Replace municipal vehicles with hybrids or those able to run off of cleaner burning diesel or E-85 fuels. Do not, however, make the change all at once. If your state or municipality has a vehicle replacement program, replace them according to that schedule and the impact will be much less. State and city officials can also encourage other public entities, such as school districts and transit authorities, to take similar action.

- Pass ordinances requiring sidewalks and bike paths. Working to make your community more pedestrian and bike friendly will make people more apt to get out of their vehicles. At the present time, too many communities do not provide adequate access to grocery stores, malls, schools, etc. If it is inconvenient or dangerous to walk or bike, people won't.

For those communities looking for other, more challenging, options they should consider the following:

- Cities ands counties should change their mind-sets as to how they make land use decisions. Instead of giving priority to the "highest and best use" they should be considering the "best sustainable use." The language we use in and titles we assign to our official documents and plans shape how we think. What if we changed our "Comprehensive Development Plans" to "Comprehensive Sustainability Plans"?

- Amend state and local building codes to require more "green" buildings. At the present time many municipalities do not even recognize the benefits of green buildings in their codes. We need to encourage energy efficient and environmentally friendly building. While building green often adds substantial initial costs, operating savings in the long run generally off-set these costs.

- Include the construction of bike paths/lanes when roads are built or widened. Only about two percent of all urban trips are made on a bike. Compare that to many European and Asian cities where bicycle trips account for 25 percent of all trips. If we build good, safe bike lanes people will use them. Financially speaking, adding bike lanes when building a new road, or widening an existing one, is the best, least expensive way to go. Bike lanes should be installed on any existing street or highway that is wide enough to do so. Since, simply from a financial standpoint, this would take several years to accomplish, a plan setting forth a priority of streets and highways to receive bike lanes should be developed and adopted by the governing body.

- Make sure traffic engineers are timing traffic signals for maximum traffic flow. The U.S. Department of Transportation has stated that the simple act of re-timing city traffic lights could save huge amounts of gasoline burned while automobiles are stopped unnecessarily by a poorly synchronized traffic light system. This will take time, but not a lot of money. It will also be something citizens appreciate. At the same time, you can replace traffic signal bulbs with more energy efficient ones.

- Provide as good a mass transit system as your city can afford. Given the number of people expected to populate our cities in the next 45 years, it is not going to be possible to build enough highways and streets to accommodate the increase in automobiles. This is especially true if we remain so dependent on this form of transportation. People want their automobiles and the ability to drive them. The problem is that we don't have the ability to build and maintain the infrastructure necessary to support millions more vehicles on our roads. We cannot afford to build this infrastructure because of the associated costs and/or the land requirements. In order for us to continue to use and enjoy the automobile, alternative forms of convenient, reliable, and affordable mass transportation must be developed that will take many vehicles off the road on any given day. Otherwise we are all going to be trapped in our cars moving at a snail's pace.

- Unless the federal government is able to, states need to adopt a pollution tax on carbon dioxide emissions. Compared to emissions caps, this type of tax would be much easier to administer and police. These taxes could then be used to fund the development of alternative forms of energy and transportation. Another option would be an additional gas tax for consumers, which would reward individuals who choose hybrid or fuel-efficient vehicles.

CHAPTER FOUR:

World Food Supply

Question:
Will we be able to feed our growing population?

Answer:
If we protect our soils!

BACKGROUND

Population projections coupled with the possible effects of global warming present us with questions as to whether we will be able to feed a world with 9 to 12 billion people in 2050. Our primary sources of nourishment are fruits/vegetables, grains, meat, and seafood. We will look at each of these groups separately, but let's begin with a discussion of our soils.

Good soil should be sacred. The most important thing that we can do to protect our ability to grow crops is to protect the arable land we have. Today, however, we don't do this. We pave over our fertile soils as our communities sprawl and we do not take adequate action to prevent soil erosion. David Pimentel, an ecologist from Cornell University, recently reported that soil from the world's croplands is being swept and washed away 10 to 40 times faster than it is being replenished. Soil erosion is destroying an amount of land the size of Indiana each and every year. In the *Journal of Environmental,*

Development, and Sustainability in 2006, he wrote that "soil erosion is second only to population growth as the biggest environmental problem the world faces. Yet, the problem, which is growing ever more critical, is being ignored, because who gets excited about dirt?"[56] He goes on to explain that erosion is a "slow and insidious process that nickels and dimes you to death. One rainstorm can wash away one millimeter of soil. It doesn't sound like much, but when you consider a hectare, it would take 13 tons of topsoil – or 20 years if left to natural processes – to replace that loss."

The experts are divided as to whether we can feed future populations. In chapter one we established that there are 36.48 billion acres of land around the globe, 11 percent, or 4.01 billion, of which is suitable for agriculture. If it takes 0.625 acres of suitable land to support one person, then we can only support today's world population unless we can put more land into food production or increase the per acre yield on the existing land. Another dynamic is that there are always uncertainties about year-to-year agricultural yields because so many factors can have a positive or negative barring on crop production. Farmers are "gamblers" each year as they "bet" on having the right combination of water, sun, and soil conditions to provide a high enough yield for them to make a living. This year, for example, farmers are faced with an unexpected problem with honey bees.

Where have all the honey bees gone?

As of this writing, something strange is happening to honey bees as entire colonies seem to have vanished. It appears that bees are leaving their hives but never returning. The queen bee and a few worker bees that remain with the hive end up starving to death. This does not bode well for the world's food supply as bees are absolutely essential in pollinating some 90 flowering crops such as blueberries, apples, almonds, cranberries and watermelons. According to a FOXNews.com © report the honey bee pollinates every third bite of food we Americans eat.[57] The honey bee is a major contributor in a $14 billion business in the United States. No one knows for sure why this happening. Some scientists have concluded that the bees' internal "navigations systems" are being interfered with and they are unable to find their way back once they leave their hives. Others cite a number of culprits including cell phones, parasites, pesticides, and genetically modified foods.

This story is only one of many examples of changes in our food supply that can affect production. On the brighter side, our history of increasing crop yields over the last forty years gives us promise of increased yields in the future. During that period we experienced the "Green Revolution," where improvements in farm equipment, fertilizers, pesticides, and genetically high-yield engineered crops allowed for doubling of the world's food supply. If the population increases as predicted, the food supply will have to double again over the next 40 years to keep pace with people's needs. We are moving from the "Green Revolution" to the "Gene Revolution" as agricultural science continues to find even higher yielding genetic crops and ones better able to adapt to changing climates and droughts. There is little reason to believe that science will stop or that new methods of improving yields will not be found. The important question is how far and for how long can we continue to increase production.

The United Nations has reported that we will be able to feed a growing population. A recent study released from the Food and Agricultural Organization (FAO) concludes that the growth in global agriculture should be more than sufficient to meet the world demand in 2050.[58] Norman Borlang, a Nobel Prize winner, Iowa farmer, and father of the "Green Revolution," however, states that there are enormous challenges ahead if we are going to ensure that nine billion people are adequately fed in environmentally sustainable ways.[59] He predicts that advances in biotechnology and scientific tools will help us meet production needs, but that we will still need to convert more land to agricultural use.

Meat Production

As societies become financially better off they tend to eat more meat and seafood. The two fastest growing economies, China and India, have followed this trend. China's meat demand is doubling every ten years and poultry consumption in India has doubled in the last five years. However, this trend is not limited to the Chinese and Indians as the demand for meat, poultry, eggs, and dairy products is increasing worldwide.

During the past few decades, consumption of meat in developing countries has grown at a rate of about five to six percent per year and consumption of milk and dairy products at three to four percent. Poultry is the fastest-growing sector worldwide; it represented 13 percent of the meat production in the 1960s compared to 28 percent today.[60] The problem with this increase in consumption is the vast amount of land required to raise and feed these animals. As more and more land is irreparably damaged by livestock, reduced amounts of land are available for crop production for both humans and animals. According to a recent article in *The Guardian*, "[t]he consensus emerging among scientists is that it will be almost impossible to feed future generations the typical western diet eaten without destroying the environment."[61] If we can reduce our dependence on meat, we could also reduce the demands on our arable land and protect our soils from overexploitation. Moving towards a more vegetarian diet will greatly reduce the strain on our vulnerable topsoil.

Seafood – The Not So Boundless Oceans?

Some of the world's top scientists and environmentalists are painting a very bleak picture of the effects of over fishing, pollution, and ocean warming on the oceans' biodiversity. International ecologists appearing in the journal *Science* predict that commercial species of saltwater fish and shellfish could be wiped out on a global scale by 2048 if present trends of over fishing continue. The report claims that 90 percent of large predator fish, such as sharks and grouper, have disappeared from the world's oceans since 1950 as a result of over fishing. [62]

The marine ecosystem, like any other, is delicately balanced. Removing one species of fish from this balance can have disastrous consequences. The over fishing and the needless killing of the great sharks has created a sharp decline in the number of Tiger Sharks, Hammerheads, and Bull Sharks. Historic records reveal that the

Tiger Shark and Scalloped Hammerheads have declined more than 97 percent since the mid-1980s, while the Smooth Hammerheads and Bull Sharks have decline by 99 percent.[63] As these species disappear, the fish that they prey on increase in numbers. For example, the Cownose Rays populating the waters along the Atlantic coast, a staple of the shark diet, have increased by 8 percent a year as the shark population declined. This in turn has let to the collapse of scallops in these waters as Cownose Rays now are left undisturbed to feast on the scallops; essentially wiping them out. Marine researchers fear that without sharks to keep them in check rays will consume shellfish in such numbers that they will be unable to recover.[64]

The collapse of many species such as scallops has been highly publicized in recent years. The following are some recent statistics on the issue:

- "A look at 1,000 years of historical data from 12 coastal regions around the world, including San Francisco Bay, showed that on average, 38 percent of economically important species had collapsed and one out of 15 had gone extinct."[65]

- A study out of Australia reported that 29 percent of the 8,000 fished species are considered "collapsed"; that is, their catches had declined by 90 percent or more.[66]

- The worldwide fishing industry currently extracts some 80 million tons of fish each year from the world's oceans. Global fish take reached 100 million tons in 1989 and increased to 109 million tons in 1994 before falling to the current levels.[67]

In 2006, the European Union received a report from the World Wildlife Fund (WWF) which warned that wasteful and illegal fishing is decimating Europe's oceans. The report stated that the basic problem is "that there are too many fishing boats," a situation encouraged by some $950 million in annual European Union subsides. It finds that for many of Europe's commercial stocks, numbers of adult fish are 10 percent of what they were three decades ago, and notes that European Union fleets have depleted fisheries in the waters of numerous other countries as well.[68] Cod, for example, is the most over fished of all the commercial fishes. Over the past 30 years, global cod catches have dropped by 70 percent.[69] Without protective laws limiting cod take, the species is in danger of simply disappearing from our tables.

But cod is only one of the many endangered fish. Swordfish and bluefin tuna are in worse shape. A report to world governments delivered in 2006 at the annual meeting of the International Commission for the Conservation of Atlantic Tunas (ICCAT), stated that "Bluefin tuna in the Eastern Atlantic and Mediterranean Sea are at such high risk of fishery and stock collapse that scientists say that the allowable catch should be halved to conserve them."[70] Unfortunately, the governments whose fleets are involved in fishing for these giant tunas did not back this recommendation and passed a much weaker plan that allows for much higher catch levels. As cod, swordfish, and bluefin tuna become scarcer, they become more valuable and therefore more profitable for fishermen to catch. As the demand for sushi and sashimi has increased, the bluefin tuna became the most valuable fish in the seas. Just one bluefin tuna, for example, could sell for tens of thousands of dollars as currently the "highest amount paid for a bluefin tuna was $180,000 on the Japanese fish market."[71]

Hope for our ocean species is not lost, however. Studies show that we "have not yet reached the point of no return. Sound fisheries management can make the difference in allowing fish stocks to recover. Disciplined fisheries management limiting the numbers of fish caught to what can be sustained is the only answer to the long-term health of the fishing industry. Extinct fish are hard to catch."[72] The impact of any regulation will not be unfelt by our cities and towns, however. Many coastal communities rely heavily on the fishing industry for the livelihood of their citizens. Local governments in these situations need to consider the impact of these changes.

One step the federal government has taken to protect our oceans, and the species living in them, is to create national protected waters. One example is the "designation of a chain of islands spanning nearly 1,400 miles of the Pacific northwest of Hawaii as a national monument created the largest protected marine reserve in the world. The area encompasses nearly 140,000 square miles, an area about the size of Montana and larger than all the national parks combined." The reserve also "supports more than 7,000 marine species, at least a fourth of which are found nowhere else on Earth. The islands included in [this] protected area include almost 70 percent of the nation's tropical, shallow-water coral reefs, a rookery for 14 million seabirds, and the last refuge for the large predatory fish at a time when 90 percent of such species have disappeared from the world's oceans."[73] Perhaps states and other local governments can work to protect our bodies of water in the same way.

Connect the Dots: Global Warming

Damage to marine species, especially those living in coral reefs, is also a result of rising ocean temperatures. Here we see the connection to global warming. Coral reefs are some of the richest marine environments in the world, but also the most fragile. One example is an area that lies between Papua-New Guinea and Indonesia known as Raja Amput. This area is "recognized as the centre of biodiversity in the world's oceans." The problem is that when the "coral reefs become stressed…by a slight rise in temperature, a change in salinity or pollution… [they become] 'bleached' and soon [die]. Ninety-four percent of the coral reefs in this area have already been damaged."[74] As was mentioned in the previous chapter, the increased levels of carbon dioxide in the atmosphere are making the world's oceans more acidic. Scientists from the National Center for Atmospheric Research and the National Oceanic and Atmospheric Administration warn that the increase in carbon dioxide is "dramatically altering ocean chemistry and other marine organisms that secrete skeletal structures."[75] Why is this important? When acidity levels rise significantly, marine life that need calcium carbonate to form their shells and skeletons are damaged. Carbon dioxide forms carbonic acid when it is absorbed by seawater, which in turn lowers the pH level making the water more acidic. This acidity makes it more difficult for marine life, such as coral, to form. Reducing our greenhouse gas emissions will not only benefit our climate, but our oceans as well. Ken Calderia, a chemical oceanographer at Stanford University, states "What we're doing in the next decade will affect our oceans for millions of years."[76]

Corn: Food or Fuel?

One alternative source of energy that the United States in particular has begun utilizing is ethanol. Fred Michel, Associate Professor at Ohio State University, and many others believe that ethanol has the potential to be a substitute for a large percentage of the gasoline we consume.[77] The concern, however, is that ethanol is produced from corn and as the demand for alternative fuel increases, so will the price of corn. In other parts of the world, corn is a diet staple for many poor countries. If people are forced to compete for corn with ethanol producers, for example, who will win? "The amount of corn needed to generate one tank of ethanol for a 25-gallon SUV gas tank would feed one person for a whole year" in many of these countries.[78]

Does this mean that supermarkets and ethanol producers will be fighting for resources in the future? Professor Michel notes that given the amount of corn we produce we could use ethanol as a substitute for approximately eight or nine percent of our gasoline used for transportation, without disturbing the food supply. But the amount of ethanol produced in the United States is growing rapidly. With the increases in gas prices and government ethanol subsidies, ethanol is becoming more profitable each year. Between October of 2005 and October 2006, 54 ethanol distilleries were constructed. When these facilities are up and running by the end of 2007, they will produce 4 billion gallons of ethanol annually, using 39 million tons of corn.[79]

It is also important to note that the water requirements to produce ethanol are significant. At the present time, it takes 300 million gallons of water to produce 100 million gallons of ethanol based fuel.[80] We will go into more detail in chapters six and seven regarding our water supply, but we cannot forget the side effects of some of these alternative energy sources.

Science to the Rescue?

Will science come to our rescue? Will we develop new food production methods in the future to feed a growing population? It is very likely that some advances will be made that will allow us to increase production, but should we rely solely on that chance? Perhaps we will develop new fishing techniques that reduce the number of unintentionally killed marine life. Or find new ways to increase fish populations. Perhaps we will develop crops better able to adapt to changing weather conditions. Or new ways to speed up fertilization of our soils. The trouble is that no one knows for sure what scientific and technological advances will be made ten or even five years from now. What we need to do is provide the leadership necessary to ensure a sustainable future. Again, we will need to incur short-term hardships to gain long-term benefits. And this is generally unpopular with the public. We would likely see some people suffer in the short run, but will the temporary pain avoid extinction for many species and ensure fertile soil for years to come?

Another Solution? – Biomass

In trying to solve one problem, we often create another one. One example of this is the development of ethanol from corn. By burning ethanol we are able to reduce the burning of other fossil fuels required for our cars and trucks. However, using corn to produce ethanol reduces the amount of corn available in the national food supply for a growing population.

On the brighter side, there may be a more feasible, "green" fuel alternative to corn based ethanol. Scientists are now experimenting with a new fuel made from agricultural wastes and other plant residues. This material is known as biomass. In the past, producing ethanol from biomass has been too expensive to generate a market. The increase in gasoline prices, however, is making this alternative a more competitive one. The United States has a vast supply of agricultural and other waste materials that could be used for fuel. Imagine in the future our vehicles being powered by corn stalks, wood chips, sawdust, wheat straw, waste paper, and even switchgrass.

The development of ethanol from biomass instead of corn has some obvious advantages including:

- Biomass ethanol reduces our use of gasoline and therefore greenhouse gas emissions, which will help us to reduce our dependence on foreign oil.
- The materials being used to produce the fuel come from our waste products, instead of from our food supply.
- The amount of waste products available to produce this ethanol is far greater than the current amount of corn available.
- Thousands of acres of marginal lands could be used to produce these energy crops, such as switchgrass.[81]

A Different Meat Source

What if we told you that from a single cell, scientists could theoretically produce enough meat to supply the entire world for one full year? And, that they could produce the meat in a method that is healthier for the environment and does not require the slaughter of livestock. What would you say to that?

Scientists at NASA and the University of Maryland have discovered that a single muscle cell from a cow or chicken can be isolated and divided into thousands of new cells. Experiments using fish tissue have already created small amounts of bitter tasting meat as part of a NASA experiment to find potential food products for long-term space travel.[82] This is not, however, a new idea. As far back as 1932, Winston Churchill predicted that "fifty years hence we shall escape the absurdity of growing a whole chicken in order to eat the breast or wing by growing these parts separately under a suitable medium." Churchill was a off in his time predictions, but his visions may still come true. Scientists believe that "tissue engineering" is advanced enough now to grow bland meat, similar to the taste of spam. However, they feel that in time they can produce tasty and textured cuts that will appeal to the average meat eater.

A Local Issue

While at first glance many of these issues may appear to be strictly national or even global, they are not. Decisions made regarding all aspects of our food supply will directly impact local governments and their residents. These issues have to do with how we use our land, how we make our livelihoods, and what our future looks like. We will address this point further in the next chapter when we discuss the food supply in the U.S.

CHAPTER FIVE:

United States Food Supply

Question:
Will the United States be able to feed approximately 120 million more Americans?

Answer:
We physically have enough land, but we will have to take better care of it!

BACKGROUND

There are no doubts that we have enough land to house another 120 million people and still not approach a density of many European and Asian countries. Presently, the "entire 'human footprint' of the United States—every building, road, grocery store, farmhouse, etc. takes up a grand total of less than five percent of the United States' land mass."[83] The important questions, however, is whether we have enough suitable agricultural and pasture land and to produce enough food to feed almost 420 million Americans? And, whether we will take the necessary steps to protect our waterways to ensure a sustainable seafood supply? If we assume for the moment that we are able to partner with the federal government to adequately protect marine populations, then we can concentrate on land requirements. So, will we be able to grow enough crops and raise enough livestock to meet our needs in the future?

In order to answer that question, we need to take a closer look at how we use land in the U.S. The United States has 2.3 billion acres of land broken down by land use as follows:

- 375 million acres are in Alaska and are unsuitable for agriculture

- 1.9 billion acres are located in the lower 48 states and consist of:
  - 66 million acres of developed land (i.e. cities and towns),
  - 73 million acres of rural residential land, and
  - 349 million acres of agricultural land (i.e. corn, soybeans, alfalfa, wheat, and vegetables)

- 747 million acres are forest

- 788 million acres are range and pasture lands[84]

Given the large amount of land not dedicated for residential and commercial purposes, we can confidently say that we can easily feed 420 million people. However, the United States is also the world's largest food exporter and other nations will rely on us to protect our soils as well. This will not be an easy task for a number of reasons.

Urban Sprawl

Just as rapidly as the U.S. population is growing, so is the amount of farm land being converted to residential and commercial land uses each year. Consider the following facts:

- Since 1970, two acres of farm land have been lost every minute.[85]

- Between 1982 an 1997, approximately 39,000 square miles, or 25 million acres, of rural land was lost to urbanization. During this same time period, the U.S. population grew by 17 percent, while urbanized land increased by 47 percent.[86]

- "Over the past 20 years, the acreage per person for new housing almost doubled, and since 1994, housing lots of 10 acres or more have accounted for 55 percent of the land developed."[87]

- If present population growth and the loss of farm land continue, the U.S. will likely cease to be a food exporter around the year 2025. At that time, all of the food we grow will be needed to feed our own population. As we are the leading exporter of food in the world, millions of people in other nations will have to look for alternative food sources.[88]

The lost farm land is not all paved over for malls, parking lots, and highways. The majority was divided up to smaller parcels and sold for housing in our sprawling suburbs. As our population continues to increase our cities and suburbs are continually expanding. While many other nations in the world are landlocked and have been forced to build up instead of out, the U.S. has more than enough land and will most likely continue to grow out. The problem is that this type of growth is not sustainable. To give the reader a better idea of the size of the loss of rural land to development the American Farm Trust has equated it to the land mass of our states. For instance:

- The loss between 1982 and 1997 of 39,000 square miles or 25 million acres of rural land is about equal to the entire land mass of Maine and New Hampshire combined.

- If this conversion rate continues, the U.S. will have lost 110 million more acres of rural countryside by 2050. This is about the combined size of Connecticut, Massachusetts, Rhode Island, Vermont, Delaware, Pennsylvania, New York, New Jersey, and Virginia.

If we allow this to happen, by 2050 we will have paved over or otherwise developed rural land equal to the land mass of the entire Northeast coast.

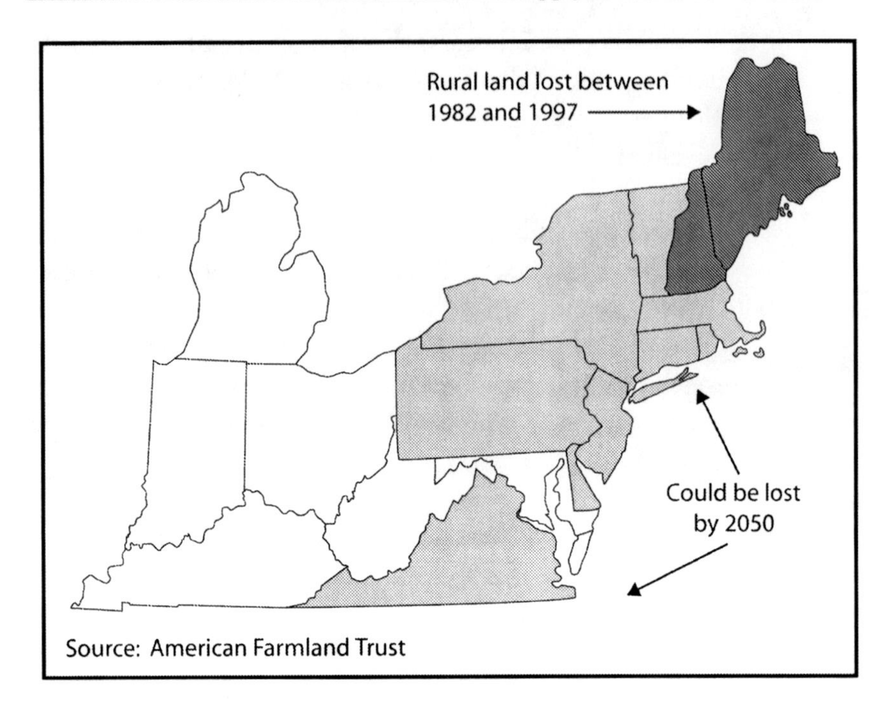

Sprawl is extremely costly and continues to eat up land and natural resources at a faster rate than we can replace them. It also makes it more difficult to provide basic services to the population.[89]

Many of our elected officials and business leaders, however, do not see urban sprawl as a threat. Rather, they see it as a sign of economic vitality. This group feels that the country's reservoir of farm land and open space is essentially limitless. They point out the fact that we produce food surpluses each year and that the government actually pays many farmers not to plant their fields each year. They argue that their push for greater economic development is required if we are to create enough jobs for a growing population. In the end, their arguments are too short sighted and negate the positive impact that open space has on communities.

Sprawl has not gone unnoticed by many people. City planners and environmental groups, for example, have pointed out the dangers of sprawl for many years. Movements have begun to try and control growth and to develop land more responsibly. Today, we all hear about new urbanism, smart growth, and urban growth boundaries. We seem to talk a great deal about the damaging effects of sprawl, but why has it been so difficult to incite change? Some of the noteworthy reasons include:

- As with population growth, the loss of farm land, forests, wetlands, and open space does not meet the CNN test. What people observe does not appear to match with what they are being told. People see large amounts of open space and farm land, especially right outside of urbanized areas. They are unable, therefore, to make the connection that there is a problem.

- Americans hate sprawl and they hate density. Solving the urban sprawl problem will depend on citizens agreeing to live in more densely populated areas. At the same time, they will have to learn to rely more on public transportation. These changes will be difficult to accept. We will need to change our view of the "American Dream" – detached houses on large lots with plenty of yard space and a two or three car garage.

- Land use decisions are made at the local level where we seldom give any thought as to how our actions will affect land sustainability. Using this strategy, the big picture is often forgotten.

Why are farms being lost?

In many areas urban sprawl would not be possible without the loss of farms across the nation. But why exactly are we losing our farms to urbanization pressure? Some of the primary reasons are:

- Pressure from Development – An increasing population will continue to drive the need for additional housing, schools, health care facilities, stores, etc. This, obviously, drives the need for additional land onto which we can build. Many developers want cheap, vacant land and local governments want increased taxes and the jobs that come from new development. Redevelopment is often much more expensive up front. As the need for this land increases, it becomes more profitable for farmers to sell their land than work it.

When the issue at the local level comes down to protecting a few acres of farm land or approving a new shopping center, the shopping center wins. The project promises new jobs and more tax dollars, so we re-zone the property and wait for the stores to open. We lose land to urban sprawl a few acres at a time, but the net effect of this slow and constant change can be significant.

- Changing Economic Sectors – Years ago the United States was a farming society, but today our economy revolves around the service sector. Farming is no longer a popular profession of choice for the majority of Americas. So, as farms are passed from one generation to the next, it becomes more attractive to many to sell their land rather than try to farm it.

- Farming Start Up Costs – The funds needed to purchase the necessary land, equipment, and materials to start a small family farming business is sizeable today.

- No Retirement Funds -- Maynard Kaufman, in an article in the Newsletter of the Michigan Organic Food and Farm Alliance, describes his dilemma as he reaches retirement age of what to do with his 160 acre farm. He believes it is in the country's best interest to preserve farm land but he can make nearly twice as much by subdividing his land it into smaller lots and selling to a developer. Since many small farmers are cash-poor and land-rich, selling their land makes economic sense for their family and helps them to secure retirement funds.[90]

- Tax and Zoning Policies -- Local governments today are still using zoning laws developed in the early 1900s, which are completely outdated. These policies make it difficult for local governments to make land use decisions that promote economic growth and, at the same time, protect the environment and our fertile soils. These zoning codes, which were originally used to separate uses as a result of the negative effects of industrial and commercial uses on residential ones, treat agricultural uses poorly. Zoning codes have historically considered all vacant or open land a temporary use. Thus, the codes did not protect agricultural land, but actually encouraged development of the land for its "highest and best use." Farm land seldom meets this definition. In addition, Americans have not come to terms with the fact that if we are to share our land with another 120 million people, we will have to allow further land restrictions. We will need regulations in place that recognize individual ownership, yet restrict property use to protect the public and our future well-being.

  Local property taxes take their toll on farmers, but not nearly as much as State and/or Federal Estate Taxes. Often, the necessity of paying estate taxes leaves no real alternative to farmers except to sell their property. Federal and state estate taxes need to be reevaluated and local government property taxes need to be restructured to protect farm ownership.

- Federal Subsidy Structure – The federal government spends approximately $20 billion a year subsidizing farmers. Six out of ten farmers, however, receive no financial assistance. Large, commercial farming operations receive 75 percent of the all subsidies and the small, family farmers are left with nothing.[91] This is hardly the path to save small farms. If we are going to continue providing farm subsidies, let us re-write the legislation to make sure that funds are spent on conservation and that financial assistance goes to the family farmers, not just the large corporations.

How can we prevent the further loss of farm land?

The federal and state governments have already developed programs to preserve land. The Department of Agriculture, for example, has a "Farmland Protection Program" that provides matching grants to local governments for the purchase of conservation easements on farm land. The "Conservation Reserve Program" is also a federally funded program administered by the National Resources Conservation Service (NRCS). This program provides farmers with technical assistance in best practices in controlling erosion, protecting wildlife habitats, water use, etc. Farmers can set aside a portion of their land for a ten-year paid lease from the NRCS.

Nineteen states have some kind of program designed to prevent the loss of farm land and open space. The most popular of these state initiatives is the purchase of development rights (PDR), which offers a less expensive option to purchasing land in fee simple. Under PDR, the State purchases only the development rights to the property. The property owner retains ownership of the land, but is restricted as to its use. If the land is sold or changes ownership through inheritance, the restrictions continue on the property and it must remain agricultural or open space forever. This is one way of allowing farmers to financially realize the increased value of their property, without selling it or changing its use.

As important as these state and federal efforts are in helping to preserve land, local governments must play a much more active role than they have to date. Why? Because a majority of land conversion decisions are made at the local level. If local governments are to be successful in preventing the further loss of land to urbanization, they must take a number of steps. They need to change their zoning practices to reflect society today, establish growth limitations, and advocate density instead of urban sprawl. Instead of concentrating on a property's "highest and best use" they should focus on its "highest and best sustainable use." Obviously our growth in population will necessitate further growth in housing and economic development, but we need to plan responsibly to accommodate these changes. If we only had to worry about our own generation and an additional generation or two into the future, allowing urban sprawl to continue would not be a problem. However, if we want to ensure a sustainable way of life for many generations to come we need to take action now.

Conclusion and Recommendations

It would appear to us that there are reasons to doubt that we will be able to feed our growing population if we continue on our current path. While we focused in this chapter on land for crops, it is just as important to remember land needs for livestock farming as we mentioned in the previous chapter. And don't forget about protecting our waterways either, as they are also essential to food production. Advances in science and technology can certainly be counted on to help us increase yields in many ways, but we can not assume they will solve the problem completely.

The Comprehensive Land Use Plan is one document that local governments can use to affect change. Many municipalities create these plans and then never look at them, or set the correct policies in place and then ignore them when it conflicts with new projects. The Comprehensive Plan is a powerful planning tool and should be built into the review process in all cities and towns. The document allows a local government to set a vision for the future and then permits them the authority to deny or grant requests according to that plan.

The most important thing that local governments can do, however, is to think bigger. Think of themselves not as an autonomous community, but as a small piece of the puzzle. Remember that small actions over long periods of time to create significant change and that local government have the power to direct our future.

CHAPTER SIX:

The World's Fresh Water

Question:
Is it possible that we have a water problem, when close to three-fourths of the Earth is covered in it?

Answer:
Yes! We have a big problem because we are running out of fresh water.

BACKGROUND

While over 70 percent of the Earth is covered in water, 97 percent of this is salt water leaving only three percent of fresh water. We are also faced with the reality that much of the fresh water is inaccessible. Our rivers are running dry and our water tables are falling on every continent.[92] Close to 70 percent of the fresh water is also frozen in ice in Antarctica and Greenland. A good portion of the remaining fresh water is present in the atmosphere at any given time and some is even in deep underground aquifers, not accessible for human use.[93] If we total all of this, we are left with less then one percent for our use.

The available fresh water is in our lakes, rivers, streams, reservoirs, and even aquifers. It is water regularly renewed by rain and snowfall; part of the constant water cycle. But why, if this cycle is constant, don't we have as much fresh water as we used to? Why do we have a problem? A number of factors influence the amount of fresh water. They include:

- Global Warming,
- Population Growth,
- Inefficient Use of Water,
- Pollution, and
- Exploitation.

We will address each of these items in this chapter. But before we start, let's look at the general impact of water shortages.

At first glance, it would appear that we have enough fresh water to support the projected world population of 9 to 12 billion people by 2050. Scientists have estimated that theoretically, there should be enough fresh water to support a population of 20 billion if it is distributed evenly around the world.[94] Unfortunately, because of weather patterns, water is not distributed evenly. In fact the countries that will experience the greatest population increases in the next 45 years are already experiencing water shortages. In 1999, the United Nations Environment Program (UNEP) reported that 200 scientists in 50 countries had identified water shortage as one of the two most worrisome problems for the new millennium.[95] Global warming ranked first among their concerns.

With the world's population growing at a rate of 80 million people a year we must find a way to add at least 64 billion cubic feet of water to the global water supply each year.[96] And this may actually be a low estimate. In the last century, water consumption grew six-fold; twice the rate of population growth. It has been predicted that if population increases occur as forecasted, and consumption patterns remain the same, two out of every three individuals will live in water-stressed conditions by 2025.[97] If we are unable to reverse this, a world crisis could develop that would affect all forms of life.

Global Warming

When global warming and water are discussed together, it is usually in the context of melting ice caps and rising sea levels. The emphasis is placed on the flooding of our major sea coast cities and low lying areas around the world, which is definitely a sobering thought. What is even more sobering is the thought of actually losing fresh water as a result of global warming. We can always move to higher ground if our oceans rise, but there is no substitute for fresh water.

The world has already been running low on fresh water in some areas. Global warming and population increases will push the supply of fresh water to the breaking point unless we take strong corrective steps immediately. Fresh water is a renewable resource, but not an infinite one. As global warming causes ice to melt, this water then spills into the oceans and becomes salient. Warmer temperatures also alter precipitation. More rain, and less snow, in the mountains, for example, will cause more runoff and less storage of fresh water. Given the large amount of impervious surface, less rain water will penetrate the ground, which means less will be available to recharge underground storage.

Snow is a natural reservoir collecting fresh water on mountains until rising spring and summer temperatures melt it. This release of fresh water comes just as more water is needed in downstream areas for irrigation. Thus as the temperatures rise with global warming and less participation falls as snow, the water runs off rather than being stored. Many of the major rivers in the world are replenished by melting snows from the mountains. As a result many of the world's major rivers, including the Colorado River, run dry before they reach the ocean.

Take China as an example. China has one-quarter of the world's population, but only six percent of its fresh water. China's rivers are supplied by water from snow in the Himalayan Mountains. Less snow, however, will mean less water to replenish China's rivers. Today, "more than 400 of China's 600 cities already suffer water shortages."[98]

Population Growth

After global warming, population growth is the factor most affecting our supply of fresh water. As the population increases, there will be less fresh water per person. Global water consumption is doubling every 20 years---or at twice the rate of population growth. By 2025, demand is expected to outstrip supply by 55 percent.[99]

As we previously mentioned, water resources are also not distributed uniformly across the globe. This also results in falling water tables in many nations. "Aquifer depletion is a new problem. Water tables are falling from the over-mining of groundwater in large portions of China, India, Iran, Mexico, the Middle East, Africa, Saudi Arabia and the United States."[100] Four hundred and fifty million people in 31 countries already face serious shortages of water, and these shortages occur almost exclusively in developing countries, which are ill-equipped to properly address them.

Inefficient Use of Water

It is absolutely possible that we can meet the world's water needs in the future by conserving and using our existing qualities more efficiently. Today, 70 percent of our total fresh water consumption is for agricultural purposes. Of the remaining 30 percent, 20 percent is used for industrial purposes and 10 percent by residential users.[101] Given that agriculture is by far the greatest user, it makes sense to work to conserve water in this area. We need to develop ways to irrigate more efficiently and reduce the amount of water lost through evaporation.

While the greatest conservation measures will have to come from better water management practices in agriculture, local governments can also help. One important way to conserve is to replace old and broken sewer and water infrastructure. The amount of water lost in these systems can significantly add up. In addition, communities can educate their residents about the need for water conservation and provide tips for ways everyone can lend a hand.

POLLUTION

Societies have often viewed oceans and waterways as their waste dumping grounds. Water has long been polluted by raw sewage and industrial chemicals, to name just a few. Many believe that the sheer size of the oceans means that this waste will just be diluted, however, water has a limited capacity to absorb pollution. Pollution primarily affects the quality of our fresh water, reducing the amount that is useable. Humans are hard on our fresh water bodies, especially rivers. A sampling follows:

- Of the 500 major rivers in the world, the Amazon River in South America and the Congo River in the sub-Saharan Africa are the healthiest because they have few industrial centers near their banks.[102]

- 80 percent of the major rivers in China are so polluted they can no longer support fish. The Yangtze River is contaminated from millions of tons of industrial waste and raw sewage being dumped into the waters everyday. [103]

- In Mexico only nine percent of the streams and rivers have water that is fit for drinking. Its underground water is equally polluted. [104]

- The Ardh Kumbh Mela, a ritual of washing away sins, was nearly cancelled this year as Hindu holy men declared that the Ganges River was too filthy for bathing. Only after the Indian government opened a dam and pumped fresh water into the Ganges did the immersion ritual take place.[105]

The result is that 95 percent of the world's cities are polluting bodies of fresh water, making the water unfit for our consumption. It has been estimated that every day, nearly two million tons of waste are dumped into water worldwide.[106] There are some nations that have taken steps to control pollution, but obviously much more action needs to be taken. Local governments can assist countries in protecting our water by monitoring the activities of residents and companies within their boundaries. Do we really want to continue to pollute this precious resource that is vital to the survival of all living creatures?

Exploitation

There is a rather disturbing trend developing that is sure to spell trouble for many countries. This trend is the purchasing of fresh water supplies by large, multi-national corporations. Water, like air, is a necessity of life. According to Fortune magazine, it is also "one of the world's great business opportunities. It promises to be to the 21st century what oil was to the 20th."[107]

Water, like oil, offers opportunities for huge profits. Urban populations are exploding and fresh water supplies are decreasing, which means that our water needs will only increase. Governments in many of the countries most in need of adequate and safe water supplies are financially unable to invest the necessary funds to build and operate the facilities that will be required. Enter the big corporations. The majority of municipal water systems (approximately 80 percent) remain publicly owned. However, during the past ten years, three global corporations have quietly taken control over supplying water to almost 300 million people worldwide.[108]

Larry West, writing in "Your Guide to Environmental Issues," reports that savvy investors and successful companies are turning water into gold. In the future, water may in fact become more valuable than oil. There are substitutes for oil, but there is no substitute for water. If the population predictions are true, demand for water could be much greater than the supply, thus creating a great financial opportunity. According to the Bloomberg World Water Index, the 11 private water utilities they track have returned 35 percent to investors every year since 2003, compared with 29 percent for oil and gas stocks and 10 percent for the Standard & Poor's 500 index.[109]

It is little wonder that investors like T. Boone Pickens and large companies, such as General Electric, are attempting to secure freshwater supplies to sell in the open market. Enron was pursuing this same route before it collapsed. Mr. Pickens owns land which sits on top of the Ogallala Aquifer, the largest aquifer in the U.S covering 175,000 square miles and running through five states. He plans on pumping water from the aquifer and selling it to cities in Texas. Cadiz, a private company in Southern California, is reportedly negotiating with multiple water districts. The company plans to pump and store water from the Colorado River in a Mojave Desert aquifer and sell it to the districts during periods of water shortages.

Without governmental restriction on the ownership and/or pricing of water, the cost will only continue to increase. In addition, when private companies own water distribution systems, local governments cannot hold them accountable for the quality of service they provide to residents. Water should belong to us all and should not be controlled by the private market whose primary goal is profit. Our survival depends on the long-term sustainability of the fresh water supply.

Maude Barlow, author of "Blue Gold," believes that turning such a precious life-giving resource as water over to private corporations, which make decisions based on ability to pay rather than on need, would be a horrible mistake. She gives the following reasons for her position:

- Privatization of water increases the cost to consumers. Customer fees have been shown to increase wherever private companies have taken control of water supply and delivery. Corporations must show a profit to their shareholders, and thus pass this responsibility on to the consumers by increasing water prices.

- Privatization promotes water monopolies and corruption. Government officials have been indicted in many countries for taking bribes from corporations seeking to obtain lucrative water rights and contracts.

- Privatization favors the highest bidder. For the purposes of profit maximization, corporations supply water to groups or industries based on ability to pay. The poor, whose need is just as great, cannot compete with the more affluent counterparts of their society.

- Privatization restricts the flow of information. With water under the control of private entities, the public loses access to information on water quality, especially testing for water-borne parasites and toxic contaminants. Complaints over water borne diseases and other contaminations have been shown to increase when water supply is privatized.[110]

Corporations and investors who are trying to privatize water supplies the world over, are getting some help from powerful entities such as the World Trade Organization, the World Bank, and the Asian Development Bank. The World Bank and the Asian Development Bank, in attempts to reduce governmental subsidies for water and assure more efficient operation of water treatment plants, often require private sector involvement in a project in order to receive bank loans or debt forgiveness. These are worthwhile goals as we waste a lot of water and when water is under priced, it encourages wasteful consumption by everyone. A more realistic pricing plan would do wonders for conservation. Many governments also cannot afford to build the facilities without outside investors. These policies, however, invariably lead to the privatization of the treatment and distribution systems.

The problem with private sector involvement is that corporations want water officially designated as a "need" not a "right." This is a very important distinction. If water is only a need, than it can be privately owned and sold in the market place to the highest bidder. If, however, water is recognized as a universal right, governments would be responsible for assuring that everyone has equal access to affordable water.[111] Water should be recognized by all governments as a fundamental, inalienable right of all, owned in public trust. The United Nations has gone on record recognizing water as a human right, but there no international laws or treaties that guarantee this right.

Fortunately, some nations have taken action in either adopting strong water protection laws or even writing it into their constitutions. South Africa, for example, indicates that water belongs to the people in its Constitution's Bill of Rights:

> *"Everyone has the right to have access to sufficient food and water, and that the State must take reasonable legislative and other measures, within its available resources, to achieve the progressive realization of these rights.*
>
> *Everyone has the right to an environment that is not harmful to their health or well-being; and to have the environment protected, for the benefit of present and future generations, through reasonable legislative and other measures that prevent pollution and ecological degradation, promote conservation; and secure sustainable development and use of natural resources; while promoting justifiable economic and social development."*[112]

The United States and other nations would do well to follow South Africa's lead in making access to affordable water a right.

The Politics of Water

Water shortage problems will only become more prevalent and more severe in the future. As this happens, tensions and even wars may well develop between nations over the allocation of fresh water. In fact, there are those who are predicting that the next wars will be fought not over oil, but water. Why?

About 40 percent of the world's population lives in 250 river basins shared by more than one country.[113] These include the following rivers: Tigris, Euphrates, Jordan, Nile, Colorado, Amur, Niger, Mekong, and Ganges. The countries that have the most oil are often the ones that have the least amount of fresh water. Thus, it is in the Middle East where population increases are causing more and more states to suffer water shortages. It is here that conflicts are likely to arise. Table 6-1 below illustrates the seriousness of the issue. Only three Middle Eastern countries faced water problems in 1955, compared to eleven in 1990. If we jump to 2025, 13 countries are projected to be in short supply.

Table 6-1: Water Stressed Countries in the Middle East[114]

1955	1990	EST. 2025
Bahrain	Bahrain	Bahrain
Jordan	Jordan	Jordan
Kuwait	Kuwait	Kuwait
	Algeria	Algeria
	Israel/Palestine	Israel/Palestine
	Qatar	Qatar
	Saudi Arabia	Saudi Arabia
	Somalia	Somalia
	Tunisia	Tunisia
	UAE	UAE
	Yemen	Yemen
		Egypt

Possible tensions/conflicts between nations are:

- Turkey and Syria or Iraq over the Tigris and Euphrates Rivers
- Israel and Palestine over the Jordan River
- Egypt and Ethiopia, Sudan, or Libya over the Nile River

Egypt, for instance, is so dependent on the Nile River that it is virtually monopolizing the water supply. Ethiopia, located at the headwaters of the Nile, is in need of additional water and has expressed an interest in tapping the Nile for some of its water, which could be a serious problem for Egypt. Water disputes will not be limited to the Middle-East and Africa either. The United States may also face disputes with both Canada and Mexico as each of these countries seeks to increase their supply.

The water levels in the Great Lakes have been dropping in recent years as many states have begun diverting water. Controversies are expected to arise as to how much water will be diverted and where it will be diverted to. One contributing factor, which may figure into any disputes between the U.S. and Canada, is the fact that a majority of the Canadian population lives close to the Great Lakes, while only a small percent of the U.S. population resides nearby. The U.S. will need to route the water supply long distances, which may not sit well with Canada.

Then there is the Colorado River, which is shared by a number of southwestern states before it enters Mexico. These states have allocated nearly all of the Colorado River's water among themselves, so that Mexico's portion of the river is almost dry. This is causing serious problems for Mexico, as its population is pumping underground water significantly faster than the aquifers can be refilled.[115]

Conclusion

Ensuring an equitable and sustainable system of fresh water resources within one's own borders is difficult enough, but not nearly as tough as when multiple nations are involved. Nations must therefore work independently and together on allocating and protecting their sources of water. China and the Caspian Nations have provided us with examples of the kind of actions necessary to protect this finite resource:

- China enacted a new law designed to reduce the number of regional fights over the Yellow River. The new law will allocate water usage by provinces and municipalities along the river's 3,395 miles. It will sanction and fine officials who do not comply with the regulations or take more water from the river than is permitted[116]

- In 2006, Azerbaijan, Iran, Kazakhstan, Russia and Turkmenistan, the five nations that border the Caspian Sea entered into a treaty among themselves agreed to reduce pollution entering the Caspian Sea, restore its environment, and allocate the water in a reasonable and sustainable manner.[117]

In chapter seven we will look further into fresh water issues facing the United States. Again, it is important to remember that our actions have consequences for other nations and vice versa. This is not a problem we can solve on our own.

CHAPTER SEVEN:

United States Fresh Water Supply

Question:
Does the United States have a problem with its fresh water supply similar to that of other nations?

Answer:
Yes, but on a much more manageable scale.

BACKGROUND

The United States is blessed with more fresh water than any other nation. Annual precipitation averages nearly 30 inches or 4,200 billion gallons per day within the lower 48 states.[118] This does not, however, mean that we do not have some serous problems that must be dealt with. We know that we have approximately the same amount of fresh water today for 300 million people, as we did 200 years ago with a population of four million.[119] We also know that despite this population increase, we are still able to supply our citizens and our companies enough water with only limited or spotted shortages.

What we need to ask ourselves is how large of a population can we serve with the resources available? Do we have enough fresh water for 420 million people? The answer is yes as long as we practice greater conservation and better management. As with many of the issues identified here, addressing global warming, reducing pollution, and protecting our open space are all on the "must do" list. There are additional solutions as well that we have not discussed.

Tensions between States

Just as tensions will development between nations over the right to fresh water, so will disagreements occur between the states within our own country.

The allocation of fresh water among the states will become increasing more heated as cities try to keep up with the water demands for their growing populations. This will not be limited to the fast growing southwest where water shortages are already a problem. Water allocation and inter-basin transfers have started to appear as problems in what has normally been considered "wet" states. South Carolina, for instance, filed a lawsuit against North Carolina in 2007 seeking to stop the cities of Concord and Kannapolis from pumping millions of gallons of water from the Catawba River. Cities in both states get their drinking water form this river. [120]

The two cities now take water out of the Catawba River and then return the treated wastewater to a river basin that is closer to them than the Catawba River. Concord and Kannapolis argue that it is too expensive to pump the wastewater back to the Catawba River. South Carolina cities who depend on the water in the Catawba for their own use disagree. They argue that there would not be a problem if the North Carolina cities returned the treated wastewater back to the Catawba River. What the courts will do with this issue is unknown at this writing. However, inter-water basin transfers are likely to become more and more of an issue between states and even between cities within the states as water needs grow.

Reducing Pollution

As the U.S. entered the nineteenth century, our cities and developing industries frequently dumped waste into streams, rivers, and lakes without concern for the consequences. It was not until 1972 that the federal government became concerned about the country's polluted waters and passed the Clean Water Act. At the time of its passage, only 36 percent of the nation's streams and lakes were safe to swim in or could support marine life.[121] Since the passage of the Clean Water Act, significant strides have been made in cleaning up U.S. waterways by federal agencies, state and local governments, non-profit organizations, and industries. However, the law's major goal, which was to eliminate the discharge of pollutants into waterways by 1985 and make all U.S. waters safe for fishing, swimming, and other uses by 1983, still has not been met. In fact, we have fallen very far from reaching this goal. It must, however, remain our goal and more effort must be made to accomplish it.

While there are sufficient laws in place to reduce water pollution substantially, strong, effective enforcement of those laws have been lacking. Consider some of the findings of a 2006 report from U.S. PIRG (Public Interest Research Groups) listed below:

- Nationally, 62 percent of all major industrial and municipal facilities discharged more pollution into U.S. waterways than their permits allow at least once during the 18-month period studied.

- Major facilities exceeding their Clean Water Act permits, on average, exceeded their permit limits by about 275 percent, or almost four times the allowed amount.

- Nationally, 436 major facilities exceeded their Clean Water Act permit limits for at least half (9 of the 18) of the monthly reporting periods between July 1, 2003, and December 31, 2004. Thirty-five facilities exceeded their Clean Water Act permits during every monthly reporting period.

- The ten U.S. states that allowed the highest percentage of major facilities to exceed their Clean Water Act permit limits at least once are Maine, West Virginia, Rhode Island, Connecticut, New York, Iowa, Ohio, New Hampshire, Utah, and the District of Columbia.

- According to EPA's Toxic Release Inventory, polluters discharged more than 221.8 million pounds of toxic chemicals into our waterways in 2003 alone.

- At least 853 billion gallons of raw sewage are dumped into U.S. waterways every year. U.S. sewer systems are aging; by 2025, sewage pollution will reach the highest levels in U.S. history without significant investment in wastewater treatment infrastructure.[122]

We cannot continue to lose clean water to pollution. We know how to stop pollution so it comes down to committing ourselves to enforcement.

Protecting Forests and Wetlands

Wetland, bogs, swamps, and marshes have long been thought of as hindrances to development in this country and thus expendable. Today, we recognize their importance for the following reasons:

- Water Filtration and Purification -- Wetlands act as giant sponges, soaking up pollutants and filtering them out before rain water enters our lakes, streams, and aquifers.

- Fish and Wildlife Habitat -- Wetlands are essential habitat for many species including ducks, geese, swans, herons, shorebirds, frogs, turtles, snakes, mink, otter, beaver, muskrat, and many others. Wetlands also provide spawning and feeding areas for fish, and homes for rare plants and insects. They are the base of several major food webs. In the Great Lakes region, a large percentage of endangered species depend on wetland habitats for their survival. Wetlands include wet meadows and open marshes, as well as forested swamplands.

The U.S. Department of Interior's Fish and Wildlife Service estimates that we have lost 53 percent of the nation's wetlands. While this is unfortunate, we must concentrate on how we can protect the

107 million acres that still exist in the United States, and begin to restore many we have lost. On Earth Day in 2004, President Bush announced a new initiative that had as its goal a "no net loss" of further wetlands and the restoration of three million acres by 2010. Since the President's challenge, we have added, protected, or improved 1.8 million acres of wetland as follows:

- 588,000 acres of added wetland,
- 563,000 acres of improved wetland, and
- 649,000 acres of newly protected wetland.[123]

This is a perfect example of the impact that local governments can have. This preservation did not take place all in one place, but was the accumulation of small amounts of acres persevered in numerous locations.

Advances in Technology

There are a number of existing technologies that have the potential to allow for better and more efficient use of our fresh water. The first is desalination, which is a procedure mankind has flirted with for a long time. Years ago, sailors used to hang sheepskins over the bow of their ships at night and then squeeze out the evaporated water they had trapped. During the mid-to-late 1900s, cities facing water shortages build desalination plants to convert ocean water to fresh water. However, this process was extremely expensive and was practical only in the world's most arid places such, as Saudi Arabia and other Middle Eastern nations.[124]

As technology has improved over the past 10 to 20 years, desalination has become more affordable, but is approximately double the cost of treating and purifying fresh water with traditional methods. None the less, as supplies of fresh water continue to dwindle, more and more coastal cities are turning to the oceans to meet their water needs. New plants are on the drawing boards, or under construction, in Massachusetts, California and Florida.

One of the downsides to desalination is the byproducts that are produced. When seawater is taken and made into fresh water, the byproduct is very high salinity water called brine. "The concentrations of the brines are usually found to be double or close to double that of natural seawater" and they oftentimes contain other chemicals as well that were used in the desalination process.[125] One issue that must be resolved in the planning of a these new desalination plants is how to properly dispose of the brine. It cannot be stored on land, because of the possibility that it might contaminate the groundwater and cause further problems. In many other plants around the world, the brine is sent through a pipe until it "meets and mixes with sea water at the shoreline," which appears to be the best solution at the present time.[126]

New irrigation practices/techniques also exist that allow farmers to limit the amount of water for agriculture. Techniques such as drip irrigation, low pressure sprinklers, and drip-walls require less water than traditional practices and also capture rainfall onsite. Farmers, and others as well, are also learning to plant more crops, land covers, and landscaping that require less water, often turning to native species for a particular region.

Conclusions and Recommendations

The issue of fresh water is not one that can be solved without the cooperation of all levels of government. First, it will take the basic guidance of the federal government to set the stage for water protection and conservation. Examples of this guidance include the following:

- Designation of fresh water as a "right," not a "need," of Americans. It is too difficult to hold local governments and private companies accountable if we do not establish this frame of mind.

- Take ownership/control of waterways that flow through more than one state to assure the quantity, quality, and proper allocation of those waters.

- Develop a national fresh water plan and require all states to develop plans to support the federal objectives.

- Begin partnering with other agencies and levels of government to enforce pollution laws.

- Continue to provide research funding to develop new technologies and better water management practices.

Generally speaking, state governments have often been more active in protecting the environment and water compared to the federal government. If we are going to make any real progress states are going to have to continue to provide this vital leadership. Some states have already included protection of air and water in their constitution, such as North Carolina:

> *It shall be the policy of this State to conserve and protect its lands and waters for the benefit of all its citizenry, and to this end it shall be a proper function of the State of North Carolina and its political subdivisions to acquire and preserve park, recreational and scenic areas, to control and limit pollution of our air and water, to control excessive noise and in every other appropriate way preserve as a part of the common heritage of this State its forests, wetlands, estuaries, beaches, historical sites, open lands, and places of beauty.*[127]

Here are some suggestions as to how states can play an even bigger role than they have in the past:

- Include the protection of air and water in their constitution, if they have not already.

- Purchase fresh water supplies, when available, for use by their citizens.

- Take ownership/control of waterways that flow through more than one county.

- Require all cities and counties to develop fresh water plans that support the state and federal plans mentioned above.

- Amend state building codes to recommend/require bathroom fixtures and sprinkler systems that are more water efficient.

- Pass state zoning guidelines that require municipalities and counties to incorporate the preservation of wetlands into their land-use planning decisions.

For those that believe this issue is not applicable at the local government level, we are here to disagree. Municipalities and counties have a major stake along with the federal and state governments and must do their part to preserve both the quantity and quality of fresh water for their residents, and to compete for economic development. Cities with abundant water supplies will be in the best position to compete for economic development in the future. Actions that can be taken at the local level are:

- Develop a fresh water protection and conservation plan with a 20 year timeline. Update the plan every five years.

- Purchase water rights when they are available. Maintain control of your water system.

- Reduce the amount of impervious surfaces when possible. Re-evaluate parking requirements and review your policies regarding "green" building materials, such as porous pavements.

- Rewrite zoning ordinances to give priority to the protection of good soils, wetlands, and forests, which all promote sustainability.

It has required the efforts of thousands of private citizens and non-profit organizations to pass federal, state and local legislation to get us where we are today. And we cannot let up. We cannot give up until all of our waterways are safe and we are certain we have a sustainable water plan. There will always be groups out there trying to overturn water protection laws or lobby to weaken enforcement efforts. Their arguments are always the same; i.e. it is too expensive, goals are impossible to achieve, it just won't work. But we can not stand for it. We must elect representatives that will look out for the common good. We need leaders among our elected officials that will legislate for the people. The burden is squarely on our shoulders to make our voices heard and choose wisely.

We also need community, religious, and business leaders willing to educate our elected officials, and the public in general, about the problem and the possible solutions. We are all busy and we rarely give a second thought to where our water comes from or what happens to it once we are finished with it. All our lives, we have turned on the faucet and had water for drinking, cooking, bathing, etc. It is therefore not surprising that we do not dwell on the issue. This has to change.

Individually, we can help do our part to affect change. We need to be more vocal with our elected officials, letting them know that we support legislation to protect and conserve water. We need to vocalize that we do not support the private ownership of water and that we believe it is a "right," not a "need."

We can also make our own homes more water efficient and find ways to re-use rain water on our property. Planting more trees is also important as they actually help to absorb pollutants from water. There is an old Chinese proverb that states, "the best time to plant a tree is 20 years ago; the next best time is today." We agree! Let's act more responsible individually and raise the general consciousness of the public.

CHAPTER EIGHT:

Alarm bells are ringing!

Question:
We will answer the call?

Answer:
Yes! The movement has begun!

A BEGINNING

In the seventeenth century, Galileo said it best when stating, "I do not feel obliged to believe that the same God who endowed us with sense, reason, and intellect has intended us to forgo their use." All across the country a massive grassroots effort has begun, heightening the public's senses about all of the issues here. When we started this book, a little over a year ago, only 16 percent of Americans felt global warming was an issue. In April 2007, a Newsweek survey found that 32 percent of citizens in this country feel it is a serious issue and that action is needed fight it! What a change!

Frustrated by the lack of action at the federal level, governors, mayors, religious leaders, business leaders, environmental organizations, college students, and ordinary individuals have taken it on themselves to start filling this leadership void. This "environmental movement" is still small, but it has doubled in just one year and is providing the stimulus that this country needs.

Let's look at examples of this type of leadership.

State Action

States have been active in passing new environmental laws, purchasing open space and farm land development rights, protecting water supplies, and insisting that the EPA enforce the federal environmental protection laws. Examples include:

- Ohio Governor Ted Strickland has promised to invest one-third of the state's bonds in developing alternative energy sources.

- California, Pennsylvania, Maryland, and New Jersey have joined seven other states in becoming "Clean Car States," passing legislation that will require substantial reductions in carbon dioxide emissions in new cars and trucks. The new rules vary from state to state, but most call for phasing over a five-year period starting in 2011.

- Arizona is in the process of setting aside nearly 700,000 acres of open space, preserving it from encroaching development.

- New England States (Vermont, Massachusetts, and Connecticut) along with New York and New Jersey joined together to begin charging power plants in 2009 for each ton of carbon dioxide they emit into the air. The funds from the fees charged to the power plants would be used for clean energy projects or conservation programs.

- Governor Arnold Schwarzenegger's "green crusade" in California is an inspiration to those who want governmental officials to take leadership roles in protecting the environment.

Local Government Initiatives

Local governments have been even more active in protecting the environment and taking steps to fight global warming compared to both the federal government and states. Examples are:

- The U.S. Conference of Mayors, led by mayors such as Mark Ruzzin of Boulder, Colorado, Dennis Walaker of Fargo, North Dakota, James Brainard of Carmel, Indiana, and Greg Nickels of Seattle, Washington, hosted a summit in the fall of 2006 on climate change, which encouraged mayors to take the lead in combating global warming.

- Over 400 Mayors have signed the "Mayor's Climate Protection Agreement," which sets goals to meet or exceed the Kyoto Protocol's greenhouse emission reductions. City mayors such as Greg Nickels (Seattle), Richard Daley (Chicago), Will Wynn (Austin), Jim Brainard (Carmel, Indiana), Jerry Abramson (Louisville, Kentucky) and Allen Joines (Winston-Salem, North Carolina) are introducing innovative solutions such as hybrid vehicle fleets, green buildings, solar powered traffic and street lights, and homes that are powered with renewable energy.

    - Last year 70 cities reported that they reduced carbon dioxide emissions by 23 million tons.[128]

    - Over 100 mayors reported that the reforms were so painless they have now set higher targets than those called for under the Kyoto standards.[129]

- Mayor Daley has pledged that Chicago will be the greenest city in North America. Toronto's Mayor, David Miller, is determined that Toronto will capture that honor. Both cities have aggressive programs and are setting examples for others to follow.

- Mayor Will Wynn's Austin, Texas is not conceding the greenest city award to anyone. The Austin Climate Protection Plan is also an ambitious one and includes the following goals:

    - Power 100 percent of city facilities with renewable energy by 2012.

    - Reduce carbon dioxide from the entire city fleet by 2020 through use of electric power and non-petroleum fuels

    - Achieve 700 megawatts in savings through energy efficiency and conservation by 2020.

    - Commit to lowest-emission technologies for any new power plants and carbon dioxide reductions on existing plants.

    - Boost energy efficiency in new homes and other buildings and require energy efficiency improvements in existing homes and buildings when they are sold.

    - Mayor Wynn is also leading a coalition of a dozen cities promoting "plug-in" hybrids. The City of Austin has pledged to purchase 600 of these plug-in hybrids for their fleet as soon as they are available.[130]

- Boulder, Colorado became the first city in the nation in which its citizens have voted an environmental tax on themselves. Residents and businesses will be charged the tax based on their electricity use.

County governments have also started to take action on their own.

- Paul Ferguson, the Chairman of the Arlington County Board of Commissioners, led the way in getting the country board to make their own pledge to help eliminate green house gasses.
    - Arlington County will buy more wind-generated electricity, give tax breaks for hybrid cars, require new public buildings to be green-certified, and hand out energy-efficient light bulbs to residents as part of a major push toward reducing greenhouse gas emissions.
    - The county has also reduced carbon dioxide and other emissions -- making its buildings more energy-efficient and adding hybrid vehicles to its fleets.[131]

Religious Organizations

Some religious leaders are also becoming environmentalists, leading movements to save the planet. Christians, Jews, and Muslims all feel a duty to become better stewards of the Earth for future generations. Examples include the following:

- Leaders within the World Council of Churches, whose membership includes Protestant, Orthodox, and Anglican Christians, have set off a campaign to mobilize their 560 million members to fight global warming.[132]
- Pope Benedict has made protection of the environment one of the goals of his papacy. He has called on every Catholic to lead more environmentally friendly lives.[133]
- Orthodox Patriarch Bartholomew, the Istanbul spiritual head of Orthodox Churches, is planning on taking a ship full of religious leaders to the Artic Circle to focus attention on global warming.

- The *Big Green Jewish Website* urges Jews into environmental protection actions.

- Shaykh Ibrahim Mogra, chairman of the Muslim Council of Britain's inter-faith committee, stated, "We believe that we are God's deputies on the planet and we have been given the responsibility to ensure we use God's gift in the correct manner and leave it in a state which can be passed on to future generations.[134]

- In San Francisco, Episcopal priest Sally Bingham formed a group called Interfaith Power and Light to assist congregations in teaching ways to conserve energy in their buildings. There are now 400 churches, synagogues, and mosques that have joined.[135]

Business Efforts

Business leaders are also realizing the threat of global warming and the financial value of going green:

- The Chief Executive Officers (CEOs) of major companies such as Alcoa, BP America, Caterpillar, Duke Energy, DuPont, General Electric, Lehman Brothers, and a host of others in January of 2007, challenged the President and Congress to pass legislation requiring mandatory reductions in greenhouse gas emissions from all emitting sources (i.e. buildings, homes and vehicles).

- Many of the top U.S. companies are developing and implementing green strategies for their own facilities and operations:

    - Wal-Mart, the world's largest retailer and the largest private user of energy, is reported to be experimenting with a green building strategy aimed at cutting costs by reducing its energy consumption by 30 percent in each of its 3,800 stores.

- Google is installing over 9,200 solar panels in its Silicon Valley complex. This represents the largest solar installation on any U.S. corporate campus to date.[136]

- Ford Motor Company converted its massive and aging River Rouge assembly plant in Dearborn, Michigan into a model "green" factory.

- General Motors unveiled a new plug-in electric car (The Volt) at the 2007 North American International Auto Show in Detroit. GM also announced that production has begun on a new generation of electric vehicles. At the same time, however, the automobile industry's powerful lobbyists continue their opposition to any attempts by congress to set higher fuel economy standards.

- Hydrogen prototypes from several auto makers are being tested as this is being written. The Chevrolet Sequel may be on the market by 2010 with Honda's hydrogen model hitting the dealers' show rooms by 2008.

- NASCAR, the country's largest racing series, has decided to race on alternative fuels during the 2007-2008 series. While the overall impact on the environment may be small, the symbolism will not go unnoticed by racing fans. Other racing series are also embracing renewable fuels. The Indianapolis 500 Race was run this year (2007) on 100 percent ethanol.

- The National Rifle Association (NRA), one of the strongest and most powerful political lobbying groups in the United States have become concerned about the loss of animal and bird habitats. They maintain that President Bush's energy and mining policies have destroyed millions of acres of land for hunting. According to a *Field and Stream* survey, 41 percent of the NRA's members feel that habitat loss is the hunter's most significant problem.

- Non-profit land trusts and conservation groups are continuing their purchases of properties and development rights, thereby saving thousands of acres of open spaces, farm land, and wetlands. However, the Bush Administration has proposed for the second year in a row the selling of approximately 300,000 acres of National Forest Lands for development.[137]

College Students

College students are organizing environmental groups on both major university and small college campuses across the country. This is a good thing considering that it will be their generation, and their children's generation, that will suffer from the negative effects of population growth, global warming, etc.

On April 12, 2007, thousands of individuals from across the country took part in over 1,300 rallies to draw attention to global warming and to push Congress and the President into passing meaningful legislation to decrease greenhouse gas emissions. Many of these rallies were led by college students and this movement will only grow. In the past, college students have proven that they can be a tremendous force for change when they are motivated and organized. Global warming has the potential to be the rallying cry that launches another major student movement in this country and throughout the rest of the world as well.

Individual Contributions

Many high-profile individuals have also joined the fight against some of these issues, including Al Gore, Bill Clinton, Tony Blair, Prince Charles, Robert Redford, Sir Richard Branson, Sheryl Crow, Oprah

Winfrey, and Tom Hanks. And the list goes on. These individuals are using their ability to draw publicity to voice their concerns for these issues and raise our consciousness. On July 7, 2007, some of the world's finest entertainers conducted "Live Earth Concerts" in seven different cities on seven different continents, to inform the citizens of the world of the threats posed by global warming. Hopefully these kinds of efforts will spur world leaders to work together to address this problem and many others.

Looking at the list of celebrities involved it is easy to question how an ordinary citizen can make a difference. But we can! We all need to do our fair share because this is not someone else's problem. This is our problem!

We don't have to install solar panels on our homes or drive electric cars to help make a difference. As individuals we contribute most of our greenhouse emissions from two sources: our homes and our cars. To the extent we can reduce our energy needs supplied by fossil fuels, we are helping to control global warming. We can not all run out and trade a perfectly good gas burning car for a hybrid. However, as we need to replace that vehicle, we can replace it with a more environmentally friendly one. Until we need to replace the car, we can set goals to drive less. According to the EPA, the average U.S. driver travels 15,000 miles per year in their cars and burns 600 gallons of gasoline. If each driver set a goal of reducing gasoline usage by 5 percent (30 gallons) we could save approximately 6 billion gallons of gas a year. Since each gallon of gasoline burned produces 5.3 pounds of carbon, this 5 percent reduction could prevent 32 billion pounds of carbon from being dumped into the atmosphere each year. This small change in our driving habits would make a big difference.

In our homes, we can reduce some of the energy needed for heating/cooling and lighting in a number of inexpensive ways such as adding insulation and using energy-efficient bulbs. Replacing one 100-watt light bulb with a 30-watt compact fluorescent bulb (which is just as bright) cuts more than 1,300 pounds of carbon dioxide pollution over the life of the bulb. Swap out two bulbs and you lower emissions by more than a ton.

Of course, the more individuals can substitute their energy needs away from fossil fuels to alternative energy sources the more beneficial it will be in slowing global warming. We all need to start taking notice of the impact we make in the homes we live in, the transportation we use, and our chosen lifestyle.

Conclusion

To be truly successful, grassroots efforts must be organized to educate and encourage elected officials at all levels to place global warming and its accompanying threats at the forefront of their agendas. We, together with our government officials, have a strong obligation to address these issues. Let us join together and make these problems major issues in all the upcoming elections. Protecting the environment will hopefully be part of every presidential candidate's platform in 2008. It is our job as citizens to react to the bells.

Chapter Nine:

A Summation

The world faces two major problems which feed off each other. It will be difficult to slow global warming without slowing population growth. It doesn't take a rocket scientist to connect the dots in how population increase and global warming are interrelated:

- The human population is growing fast.
  - o More people will need more food, more fresh water, and more shelter.
  - o Good agricultural lands and climatic conditions will be needed to grow crops and raise live stock to feed another 3 billion people world wide, 119 million more in the United States, while many acres are still being lost to erosion, poor farming practices and urban sprawl.
  - o The desire for seafood is causing over-fishing and over-harvesting of both saltwater and freshwater species to an unsustainable level.
  - o Heating and cooling our buildings and powering our automobiles with fossil fuels creates greenhouse emissions into the atmosphere.
  - o Greenhouse emissions fuel global warming

- Global warming, if allowed to go unchecked will result in:

    - o Rising sea levels, flooding many low-lying areas and many of the coastal cities.
    - o Flooding of cities which will cause mass migrations of people inland, requiring more land to be used for development.
    - o Changes in the amount, frequency and pattern of precipitation
    - o Increased desert areas due to droughts
    - o Glacial melting
    - o Reduced summer stream flows
    - o Animal and plant extinctions
    - o Increase in some diseases
    - o Warming of ocean waters resulting in destruction of coral reefs.

- Inaction:

    The time to take action is now; tomorrow may be too late. We need political, business, religious and individual leaders to recognize the threats we face and set a new course that will safely navigate us through these troubled waters of population expansion and global warming.

    However, if the past is any indication of the future, our hope lies not in our politicians. Our real hope lies with us. Every year that we continue along the path we are on is a year lost in solving these massive problems.

Quick action:

The Earth is warming and while we humans can slow the rise in temperatures, it is doubtful that we will be able to stop the warming trend. We can, by taking action now, slow down global warming to a manageable problem that will provide humans, animals and plants many more years to adapt or advance technology to deal effectively with Earth's changes. If we do nothing, we will continue in motion events that could have far-reaching and adverse consequences for our children and grandchildren.

Chapter Ten:

The Final Word

Over the course of the next 45 years, communities around the world will see the effects of immense population growth. This growth will make it more and more difficult to provide basic services to our residents, especially given the other major issues that are connected to it. Local governments are closest to the people, and will therefore be impacted more than any other level of government. When the public wants answers relating to any issue that affects their daily lives they turn to their local officials. They will be far more likely to walk into their local board/council meeting than to contact their state or federal representatives.

But the goal of this book is not to create fear. We live in a society often driven by fear of all kinds. While some of our findings were frightening to think about at times, we still feel extremely hopeful about the future. We wrote this book to incite action. Local governments across the country need strong leadership. They need leaders who are unafraid of facing these issues. They need leaders that will force discussions locally on how to address the many uncertainties we will face in the coming years.

These leaders need to remember that while we have separated issues facing local governments in this chapter, all of the issues are intertwined. It is hard to discuss effective and affordable public transportation without talking about a reduction in urban sprawl, or global warming without the effects of population growth. They are all individual pieces to a puzzle that needs to be put together.

While some of these issues are beyond the scope of many local governments, small efforts in each community do matter. The federal government has not begun to address many of these issues so it is up to us at the local level to motivate our citizens. It is easiest for most people to participate at the local level and they will be looking to mayors, city councils and city managers for answers in the future. Change oftentimes begins small and then grows.

Service Provision

Why is it that the impact of this population growth will be especially difficult for local governments to deal with? Local governments primarily exist to provide services to their residents. They pick-up trash, pump water to homes and businesses, provide protection, put out fires. What local governments do is dictated by those they serve and as that population changes and grows so will they. They are closest to the people and will have to learn to adapt more quickly than others. If they don't, they will have residents at their doorsteps demanding change.

All of the major issues discussed so far will impact each area of service differently. Our goal is to make you think about ways to adapt given a world that will constantly be changing around you. Have you sat down to think about how your departments will have to react to these changes? How will your Health and Human Services Department change their social service and senior programs? How will people get around in your community? If population densities and the elderly population increase how will those changes alter the role of your Fire Department? The questions are never ending, but need to be asked.

WHAT IF WE ARE WRONG?

What if we make all of the sacrifices and we are wrong? What if the Earth is not warming enough to cause the disastrous consequences that scientists are forecasting? We may never know because we will not know for sure what we prevented.

The authors feel that, even without the threat of global warming, we should be making the changes suggested in this little book. It would appear to us if we start now in making the changes suggested to slow both population growth and global warming, we will end up with a better world for all generations to follow. It is hard to see how we can go wrong by:

- Protecting our arable land and food supplies
- Protecting our fresh water supplies
- Protecting our clean air
- Protecting animal, bird and aquatic life
- Positioning the country to be less dependent on fossil fuels and thus less dependent on foreign governments.
- Developing an energy self-sufficiency program that makes the United States more energy secure while improving our balance-of-trade deficit by buying less foreign oil.
- Adopting an immigration policy based on defining our country's needs and recruiting the talent needed for us to stay competitive in a global economy.

These all seem reason enough to call on all of us to make some sacrifices in our lives; however, the biggest reason of all is that the stakes are simply too high to gamble on the scientists being wrong. Given the magnitude of the future threats we are dealing with this would be a very risky position to take. The stakes are high and there is no room for error. If we err, should we not err on the side of taking the safest course?

Earth is a lovely little planet. Are you willing to help save it?

References

Chapter 1

[1] David, Leonard. "World population hits 6.5 billion – Rapid growth occurring where it can be least afforded, researchers say." *LiveScience.* February 25, 2006. Available at www.msnbc.com, accessed on October 13, 2006.

[2] Malthus, Thomas. *An Essay on the Principle of Population.* 1798.

[3] Ehrlich, Paul R. *The Population Bomb.* New York: Ballantine Books, 1971.

[4] Colliers Encyclopedia.

[5] United Nations. FAO Report. Jan 1992.

[6] Wikipedia Encyclopedia – Search "Tsunami." Available at www.wikipedia.org, accessed on November 10, 2006.

[7] Reuters Foundation – AlertNet. May 26, 2006. Available at www.altertnet.org, accessed on May 26, 2006.

[8] Nielsen, Ron. *The Little Green Handbook: Seven Trends Shaping the Future of Our Planet.* New York: Picador, 2006.

[9] Nielsen, Ron. *The Little Green Handbook: Seven Trends Shaping the Future of Our Planet.* New York: Picador, 2006.

[10] Kupchinsky, Roman. "Russia: Tackling the Demographic Crisis." *Radio Free Europe/ Radio Liberty.* May 19, 2006. Available at www.rferl.org, accessed on May 26, 2006.

[11] United Nations – Department of Economic and Social Affairs: Population Division. "World Population Ageing: 1950 – 2050." 2002. Available at http://www.un.org/esa/population/publications/ worldageing19502050/, accessed on May 12, 2006.

[12] David, Leonard. "World population hits 6.5 billion – Rapid growth occurring where it can be least afforded, researchers say." *LiveScience.* February 25, 2006. Available at www.msnbc.com, accessed on October 13, 2006.

[13] Lederer, Edith. "U.N.: Rich Nations Seek More Workers." *Houston Chronicle.* April 2, 2006. Available at www.houstonchronicle.com, accessed on December 16, 2007.

[14] Lederer, Edith. "U.N.: Rich Nations Seek More Workers." *Houston Chronicle.* April 2, 2006. Available at www.houstonchronicle.com, accessed on December 16, 2007.

[15] United Nations Human Settlements Programme. "State of the World's Cities 2006/7." 2006 (http://hq.unhabitat.org).

[16] Wikipedia Encyclopedia – Search "Kenneth Boulding." Available at www.wikipedia.org, accessed on May 11, 2006.

[17] Professor Bruce Bridgeman

[18] Longman, Phillip. *The Empty Cradle: How Falling Birth Rates Threaten World Prosperity and What to Do About It.* Massachusetts: Basic Books, 2004.

[19] Rutsch, Horst. "'Literacy as Freedom' – United Nations Launches Literacy Decade (2003 – 2012)." *United Nations Chronicle.* February 21, 2003. Available at http://www.un.org/Pubs/chronicle/2003/webArticles/022103_literacy.html, accessed on May 13, 2006.

[20] Fursenko, Audrey. "Outside View: G*'s War on Illiteracy." *United Press International.* June 23, 2006. Available at http://www.upi.com/InternationalIntelligence/view.php?StoryID=20060623-010652-9324r, accessed on June 24, 2006.

Chapter Two

[21] U.S. Census Bureau. 2000 Census. As reported by *Negative Population Growth.* Available at www.npg.org, accessed on April 28, 2006.

[22] U.S. Census Bureau. Available at www.census.gov.

[23] U.S. Census Bureau. Available at www.census.gov.

[24] U.S. Census Bureau. Available at www.census.gov.

[25] U.S. Census Bureau. 2000 Census. Available at www.census.gov.

[26] U.S. Census Bureau. Available at www.census.gov.

[27] U.S. Census Bureau. Available at www.census.gov.

[28] Population Reference Bureau. "Interview with Linda Jacobsen on the U.S. at 300 Million: Challenges and Prospects." October 11, 2006. Available at http://discuss.prb.org/content/interview/detail/673/, accessed on April 30, 2006.

[29] Population Reference Bureau. "Interview with Linda Jacobsen on the U.S. at 300 Million: Challenges and Prospects." October 11, 2006. Available at

http://discuss.prb.org/content/interview/detail/673/, accessed on April 30, 2006.

[30] Population Reference Bureau. "Interview with Linda Jacobsen on the U.S. at 300 Million: Challenges and Prospects." October 11, 2006. Available at http://discuss.prb.org/content/interview/detail/673/, accessed on April 30, 2006.

[31] Negative Population Growth. "Historical and Future Population Trends." 2006. Available at www.npg.org/popfacts.htm, accessed on June 11, 2006.

[32] U.S. Census Bureau. Available at www.census.gov.

[33] Haines, Michael, as reported by Jonathan Last. "One Last Thing: The Population Contraction." The Philadelphia Inquirer. May 21, 2006. Available at www.philly.com, accessed on May 27, 2006.

[34] "People and Ethnicity: The face of Our population," U.S. Census Bureau, American FactFinder, http://factfindere.census.gov, 2-28-07

[35]. Negative Population Growth. "Historical and Future Population Trends." 2006. Available at www.npg.org/popfacts.htm, accessed on June 11, 2006

[36] Negative Population Growth. "Historical and Future Population Trends." 2006. Available at www.npg.org/popfacts.htm, accessed on June 11, 2006.

Chapter Three

[37] Achenbach, Joel. "Hurricane expert Gray, other scientists: Talk about global warming is a hoax." *Wilmington Star-News*. May 31, 2006. Available at www.www.wilmingtonstar.com, accessed on June 14, 2006.

[38] Lindzen, Richard. "Why so Gloomy?" Newsweek International. April 16, 2007. Available at http://www.msnbc.msn.com/id/17997788/site/newsweek/, accessed on April 30, 2007.

[39] Borenstein, Seth. Associated Press. "Scientists say Artic core truth is North Pole was once tropical." Printed in the Wilmington Star-News. June 1, 2006.

[40] Hansen, Jim. "The Threat to the Planet." *The New York Review.* June 16, 2006. p. 12-16.

[41] Encarta Encyclopedia.

[42] Cicerone, Ralph. "Finding Climate Change and Being Useful." Sixth Annual John H. Chafee Memorial Lecture on Science and the Environment. Lecture given January 26, 2006 at the Ronald Reagan Building and International Trade Center in Washington D.C. Sponsored by the National Council for Science and the Environment.

[43] Cicerone, Ralph. "Finding Climate Change and Being Useful." Sixth Annual John H. Chafee Memorial Lecture on Science and the Environment. Lecture given January 26, 2006 at the Ronald Reagan Building and International Trade Center in Washington D.C. Sponsored by the National Council for Science and the Environment.

[44] Encarta Encyclopedia.

[45] World Resources Institute and California Energy Commission, printed in USA Today.

[46] Hansen, Jim. "The Threat to the Planet." *The New York Review.* June 16, 2006. p. 13.

[47] Flavin, Chris. Quoted in *Environment News Service.* "Earth's Vital Signs Under Burden of Human Pressure." July 27, 2006.

[48] "Noted", *The Week.* June 9, 2006. p. 18.

[49] Crane, David. Toronto Star. April 21, 2006. Available at www.thestar.com.

[50] *The Week,* reporting on a June 2006 article in *The Washington Post,* 6-9-06, p.18

[51] George Gedda, "Inuits Claim U.S. Emissions Violate Human Rights," Anchorage News, March 4, 2007

[52] Wikipedia Encyopedia, "Global warming," http://en.wikipedia.org/wiki/global_warming

[53] John M. Broderd and majority Connelly, "Public Remains Split on Response to Warming," www.nytimes.com, 4-27-07

[54] *Financial Times,* April 25, 2007, posted by Karim Ahmed, www.earthportal.org

[55] "Engineering a Cooler Planet," www.msnbc.com, 4-2-07

Chapter Four

[56] Pimental, David. "Increasing Soil Erosion Threatens World's Food Supply." *Journal of the Environment, Development and Sustainability.* March, 23, 2006.

[57] " Bee Colony Collapses Could Threaten U.S. Food Supply," FOXNews.com, 5-03-07 and also, Mike Leonard, "Feeling the Sting: Colony collapse disorder" Hoosier Times, April 1, 2007, p.F6

[58] United Nations FAO. Economic and Social Department. *Technical Interim Report.* April 2000. p. 249.

[59] Finfacts Team. "Enormous tasks ahead to feed the world, says former Nobel Prize recipient." May 26, 2006.

[60] United Nations FAO. Economic and Social Department. *Technical Interim Report.* April 2000. p. 249.

[61] Sample, Ian. "Death of a predator: big sharks are disappearing- and world's fisheries are suffering as a result." *The Guardian.* April 22, 2007. Available at www.guardian.co.uk/science, accessed on April 24, 2007.

[62] Journal Science story reported by David Helvarg and David Helvarg, "Too few Fish in the Sea," www.latimes.com, 11-8-06

[63] Sample, Ian. "Death of a predator: big sharks are disappearing- and world's fisheries are suffering as a result." *The Guardian.* April 22, 2007. Available at www.guardian.co.uk/science, accessed on April 24, 2007.

[64] Sample, Ian. "Death of a predator: big sharks are disappearing- and world's fisheries are suffering as a result." *The Guardian.* April 22, 2007. Available at www.guardian.co.uk/science, accessed on April 24, 2007.

[65] Mason, Betsy. "Seafood supply to dry up by 2048, study says." *ANG Newspapers.* November 3, 2006. Available at www.insidebayarea.com, accessed on November 5, 2006.

[66] Reported in "Governments doubt dire fishing threat." June 11, 2006. Available at www.abc.net.au.

[67] Green Peace. "Fisheries and the World Food Supply." Available at http://archive.greenpeace.org/comms/fish/am02.html, accessed on November 9, 2006.

[68] "Unsustainable Fishing Threatens Europe's Oceans." September 29, 2006. Available at www.idn-newservice.com.

[69] "Unsustainable Fishing Threatens Europe's Oceans." September 29, 2006. Available at www.idn-newservice.com.

[70] *Environment News Service.* "Fishing Nations Split Over Endangered Bluefin Tuna Conservation." December 1, 2006. Available at www.ens-newswire.com.

[71] *Environment News Service.* "Fishing Nations Split Over Endangered Bluefin Tuna Conservation." December 1, 2006. Available at www.ens-newswire.com.

[72] "No Fish Story: Imagine a World Without Seafood," Washington-Post, 11-20-06, P.A16, www.washingtonpost.com

[73] Eilperin, Juliet. "More carbon dioxide causing higher level of acidity in oceans." *The Washington Post.* June 25, 2006. p. 5A.

[74] Blundell, Nigel, "Meltdown." Sunday Mirror, www.sundaymirror.co.uk., November 15, 2006.

[75] Eilperin, Juliet. "More carbon dioxide causing higher level of acidity in oceans." *The Washington Post.* June 25, 2006. p. 5A.

[76] Eilperin, Juliet. "More carbon dioxide causing higher level of acidity in oceans." *The Washington Post.* June 25, 2006. p. 5A.

[77] Kelly, G. Patrick. "Panel talks about alternative fuel sources." *The Times-Reporter.* December 1, 2006. Available at www.timesreporter.com, accessed on December 3, 2006.

[78] Brown, Lester. Quoted in the KCCI, Des Moines News Cast. November 18, 2006. Available at www.kcci.com/news, accessed on December 14, 2006.

[79] November 7, 2006. www.ewire.com.

[80] Paul, Jim. "Ethanol: A cure or a new problem." Wilmington Star-News. June 25, 2006. p. 1E.

[81] Gillis, Justin. "A New Fuel Source Grows on the Prairie." *The Washington Post.* June 22, 2006. p. A.1. Available at www.washingtonpost.com, accessed on June 2, 2007.

[82] Sample, Ian. "When meat is not murder." *The Guardian.* August 13, 2005. Available at www.guardian.co.uk, accessed on June 2, 2007.

Chapter Five

[83] Dr. Ronald Utt of the Heritage Foundation in Washington, D.C.

[84] Colliers Encyclopedia.

[85] American Farmland Trust. Information Available at www.numbersusa.com/interests/farmland, accessed on January 11, 2007.

[86] National Resources Conservation Service. U.S. Department of Agriculture. "Smart Growth, Open Space & Farm Land." Available at www.grinningplanet.com/2005, accessed on April 17, 2006.

[87] National Resources Conservation Service. U.S. Department of Agriculture. "Smart Growth, Open Space & Farm Land." Available at www.grinningplanet.com/2005, accessed on April 17, 2006.

[88] Pimental, David. "Food, Land, Population and the U.S. Economy." *Carrying Capacity Network.* Washington, D.C.

American Farmland Trust, www.numbersusa.com/interest/farmland.html, 12-16-06]

[89] Information available at www.sprawlcity.org.

[90] Kaufman, Maynard. "Cheap Food and the Loss of Farmland: A Farmer's Perspective." *Newsletter of the Michigan Organic Food and Farm Alliance.* November/December 1997.

[91] Young, Rebecca. April 14, 2007. Available at www.care2.com.

[92] "Water Shortages." Web of Creation. Available at www.webofcreation.org/earth%20problems/water.htm.

[93] Tilford, David. "Draining the Blue Planet." December 14, 2006. Available at www.newdream.org, accessed on December 15, 2006.

[94] Illueca, Jorge and Walter Rast. "Precious, finite and irreplaceable." December 15, 2007. Available at www.ourplanet.com, accessed on December 15, 2007.

[95] Kirby, Alex. "Dawn of a thirsty century." BBC News. December 15, 2006. Available at http://news.bbc.co.uk, accessed on December 15, 2006.

[96] Larson, Samuel L. "Lack of Freshwater Throughout the World," The water Web Site, www.freshwater.org/water_facts.html, December 15, 2006.

[97] United Nations Environmental Program (UNEP)

[98] "The situation will only get worse. Wilmington Star News. July 3, 2006. p. 8A.

[99] Tilford, David. Facts complied by Tony Clark (Polaris Institute), Margaret Bowman (American Rivers), and Hans Schoepfin (Panta Rhea Foundation) at May 2003 Water Teleconference.

[100] Web of Creation. "Water Shortages." Available at www.webofcreation.org.

[101] Alex Kirby, "Why world's taps are running dry," www.bbcnews.co.uk, 6-20-03

[102] World Commission on Water for the 21st Century. "Rivers of the World Mismanaged, Polluted." January 31, 2007.

[103] Ibid

[104] Ibid

[105] The Week. "Ganges is impure." January 19, 2007. p. 7.

[106] West, Larry. "Water More Valuable than Oil?" Bloomberg World Water Index. December 15, 2006. Available at http://environment.about.com, accessed on December 16, 2006.

[107] West, Larry. "Water More Valuable than Oil?" Bloomberg World Water Index. December 15, 2006. Available at http://environment.about.com, accessed on December 16, 2006.

[108] Barlow, Maude and Tony Clarke. "Blue Gold - Fight to Stop the Corporate Theft of the World's Water." As reported by Web of Creation. August 9, 2006. Available at www.webofcreation.org/Earth%20Problems, accessed on May 13, 2007.

[109] West, Larry. "Water More Valuable than Oil?" Bloomberg World Water Index. December 15, 2006. Available at http://environment.about.com, accessed on December 16, 2006.

[110] Barlow, Maude. "Blue Gold - The Global Water Crisis and the Commodification of the World's Water Supply." as reported by Robert Svadlenka. December 15, 2006. Available at www.worldhungeryear.com, accessed on January 4, 2006.

[111] Barlow, Maude and Tony Clarke. "Blue Gold - Fight to Stop the Corporate Theft of the World's Water." As reported by Web of Creation. August 9, 2006. Available at www.webofcreation.org/Earth%20Problems, accessed on May 13, 2007.

[112] *Environment Matters.* Annual Report, 2003. p. 11.

[113] Vaknin, Sam. "The Emerging Water Wars." April 29, 2007. Available at www.opednews.com, accessed on May 13, 2007.

[114] Darwish, Adel. Lecture at the Geneva Conference on Environment and Quality of Life. June 1994. Available at www.mideastnews.com, accessed on May 13, 2007.

[115] Vaknin, Sam. "The Emerging Water Wars." April 29, 2007. Available at www.opednews.com, accessed on May 13, 2007.

[116] Vaknin, Sam. "The Emerging Water Wars." April 29, 2007. Available at www.opednews.com, accessed on May 13, 2007.

[117] Environment News Service. "Five Caspian Nations Ready to Reverse Conservation Crisis." July 27, 2006. Available at www.ens-newswire.com, accessed on October 12, 2006.

Chapter Seven

[118] Frederick, Kenneth D. "America's Water Supply: Status and Prospects for the Future." *Consequences.* Vol. 1, No. 1, Spring 1995.

[119] Frederick, Kenneth D. "America's Water Supply: Status and Prospects for the Future." *Consequences.* Vol. 1, No. 1, Spring 1995.

[120] Meg Kinnard, "S. Carolina files suit against N.C. over Catawaba River water," Wilmington Star-News, 6-8-07

[121] Sierra Club. "Clean Water." January 20, 2007. Available at www.sierraclub.org, accessed on February 8, 2007.

[122] U.S. PIRG. "Troubled Waters: An Analysis of Clean Water Act Compliance." 2006.

[123] Council on Environmental Quality. "Conserving America's Wetlands 2006: Two Years of Progress Implementing the President's Goals." April 2006]

[124] Daley, Beth. "Thirsty towns eye desalting plants." *Boston Globe.* April 10, 2004. Available at www.boston.com, accessed on April 28, 2007.

[125] Einav, R., Harussi, K., Perry, D. "The footprint of the desalination process on the environment." *Desalination.* 2002. 152:141-154.

[126] Helly, J.J., Herbinson, K.T. "Visualization of the salinity plume from a coastal ocean water desalination plant." *Water Environment Resource.* 1994. 66:753-760.

[127] North Carolina Constitution, Article XIV, Section5.

Chapter Eight

[128] Simon, Stephanie. "Global warming, local initiatives." L.A. Times. December 11, 2006. Available at www.latimes.com, accessed on January 13, 2007.

[129] Simon, Stephanie. "Global warming, local initiatives." L.A. Times. December 11, 2006. Available at www.latimes.com, accessed on January 13, 2007.

[130] McManus, Reed. "Hybrid Helpers." *Sierra Club.* March/April 2007. Vol. 92, No. 2, p. 16.

[131] Gowen, Annie. "Arlington Takes On Global Warming." *Washington Post.* January 2, 2007. p. A01.

[132] Majendie, Paul. Reuters. "World's Churches Go Green and Rally to Cause." *Environmental News Network.* February 9, 2007.

[133] Majendie, Paul. Reuters. "World's Churches Go Green and Rally to Cause." *Environmental News Network.* February 9, 2007.

[134] Majendie, Paul. Reuters. "World's Churches Go Green and Rally to Cause." *Environmental News Network.* February 9, 2007.

[135] Miller, Lisa. "God is Green." *Newsweek.* April 30, 2007. p. 12.

[136] Sierra Club. "Grapevine." March/April 2007. p. 32.

[137] President Bush's FY 2008 Budget. p. 177. Available at www.whitehouse.gov, accessed on June 8, 2007.

Printed in the United States
88423LV00005B/301-312/A

9 781434 323491